Make the Future Work

Appropriate Technology: a Teachers' Guide

edited by Catherine Budgett-Meakin

LONGMAN

Longman Group UK Limited,
Longman House, Burnt Mill, Harlow,
Essex CM20 2JE, England
and Associated Companies throughout the world.

First published 1992

Prepared for publication by
Jenny Lee Publishing Services,
Bishop's Stortford, Herts;
designed by Gillian Glen.
Set in Palatino 10/12 point by AlphaSet,
High Wych, Harlow, Essex.

Produced by Longman Singapore Publishers Pte Ltd
Printed in Singapore

The Publisher's policy is to use paper manufactured from
sustainable forests.

ISBN 0 582 08838 0

Contents

Preface

Amongst the many new tasks that the National Curriculum requires from schools, the implementation of Technology is one of the most demanding. A radically new definition has been adopted for this subject. This novelty means that teachers have to work together in new ways and with a new approach. This book is timely because it offers substantial help with this task.

However, its outstanding contribution is to make clear the importance of the new subject. Design and Technology must be concerned with choosing what we want to build and make, with deciding how – or whether – we want new technical possibilities to transform our culture. A recurring theme in these chapters is that issues of value and belief are central to technology.

Finally, there is a particular and essential emphasis here on the need for a global view of technology. Valid criteria for technological change in the future must include an audit of global effects, and in particular on the balance between the prosperous and the needy societies of the world. Not only is the appropriate technology of the developing world an interesting subject of study: it may also contain a vision of approaches that we must all adopt if this planet is to survive as a place able to support human life.

It is desperately important that education for the twenty-first century helps pupils to be aware of these issues. Because this book makes them clear, stresses their urgency, and gives practical help in meeting them, I am happy to recommend it most strongly.

Professor Paul Black
Centre for Educational Studies
King's College, London

September 1991

Acknowledgements

First and foremost I should like to acknowledge the cooperation of all the authors who have contributed chapters to *Make the Future Work*. There is a certain inherent complexity in managing a project with so many contributors (fourteen), and they have eased my task. It has been a pleasure to work with them.

I believe that the book has a vision and a cohesion, and all the authors have contributed their own perspective to that vision. It is interesting that there are instances where similar examples are used by different authors, but with different emphases. Rather than seeing that as repetition, I feel that the result provides reinforcement and strengthening of the argument.

I should also like to thank my friends and colleagues at Intermediate Technology, who have provided advice and support, and in some cases read drafts. I should also like to pay tribute to the lucky star which brought me to work for IT in 1983, and to George McRobie who smoothed my path at the outset.

Last, but not least, I would like to thank my husband, John Mead, for his encouragement, both moral and practical. It is he who has given me confidence to tackle the issues covered in Chapter 1 – he is the 'greenest' person I know, and sets me 'life-style' standards for which to aim.

Catherine Budgett-Meakin
Balham, London
September 1991

(For text and photo acknowledgements see page 191.)

Dedication

To my parents Denzil and Kathleen Budgett-Meakin, and to all those who, like them, have at heart the humanity and values which permeate this book.

Biographical details of authors

Catherine Budgett-Meakin is now Head of the Education Office for IT, Catherine is a Sociology graduate from the University of Kent. After working in the Marketing Division of Unilever in London, she crossed Africa in a Bedford truck, then taught at the University of the Witwatersrand in Johannesburg. She returned to do a PGCE at Oxford, then taught in ILEA comprehensive schools in Brixton, Tooting and Lewisham in London over a period of fifteen years. She specialised in teaching children with learning difficulties, throughout the secondary age range. In 1983 she began to work for Intermediate Technology as a volunteer, and felt that she could contribute something to IT. She and the team of four in the IT Education Office are now engaged in meeting the needs of the National Curriculum, using their specialist IT knowledge, providing INSET support, and publishing materials.

Colin Mulberg is Education Officer at the Victoria and Albert Museum, London, with overall responsibility for Design and Technology within the Schools Section. He has devised training courses to demonstrate the need to integrate social and cultural factors into school technology teaching, and has written numerous chapters and articles on this topic. He has been a long-standing adviser to Intermediate Technology, and is schools representative on the Education and Training Committee of the Institution of Engineering Designers.

David Layton (Emeritus Professor, OBE) was Professor of Science Education at the University of Leeds from 1973 to 1989. He was co-director in the national evaluation of the TVEI Curriculum from 1985 to 1988. He was a member of the Secretary of State's National Curriculum Working Group on Design and Technology, 1988–89. Author of numerous books and articles on science and technology education, including (with Gary McCulloch and Edgar Jenkins) *Technological Revolution? The Politics of School Science and Technology since 1945*. Editor of the Unesco series *Innovations in Science and Technology Education*, volumes 1,2, 3 and 4 (1986).

Ken Webster taught Economics for seven years in North Wales before joining the Manchester University Economics Education 14–16 Project. Now freelance, he is involved with in-service work on cross-curricular themes, especially environmental education, here and in Eastern Europe. He is working with WWF(UK) on the production of resources for secondary schools. His books include *Choosing the Future* (co-author), *Energy* and *The Stimulus Video Pack*.

Alan Dyson is a Lecturer in Education in the School of Education at the University of Newcastle upon Tyne. Much of his work consists of school and LEA-based in-service with experienced teachers, and he has acted as consultant to many schools in their development of policy and of curriculum projects. He previously taught for thirteen years in both comprehensive and special schools.

Roger Standen studied Art and Design before working in a range of secondary schools. As Head of the Creative Studies Faculty at a Portsmouth comprehensive school he was able to work in all the 'traditional' design areas, Art and Design, CDT, Home Economics, developing a primary link course. In 1972 he was seconded and studied the cross-curricular nature of design. From 1980 to 1982 Roger was a Research Fellow at the Royal College of Art and a member of the Schools Council and the Built Environment Project team. From 1982 he became General Inspector in the London Borough of Bromley with responsibility for the development of Art and Design, Home Economics, CDT and Business Education. Since 1985 he has directed the work of the Design Dimension Project and is now the Director of the Design Dimension Educational Trust which has its headquarters at Dean Clough in Halifax, West Yorkshire.

Sue Greig is a freelance writer, editor and facilitator in global education. From a background in science teaching and ecological research, she spent three years at the Centre for Global Education, York University, working on links between environmental and development education and on the process of organic change in schools. Together with Graham Pike and David Selby she co-authored *Earthrights: education as if the planet really mattered* (1987) and *Greenprints for Changing Schools* (1989), both published by WWF and Kogan Page. More recently she has written an activity-based teaching pack, *In the Shadow of the City* for the Save the Children Fund, and has contributed to an environmental action pack for Friends of the Earth, published by Hobsons in late 1991.

Ann MacGarry qualified as a History teacher then, among other things, taught English in Kenya. She then re-trained in order to teach Design and Technology. After eight years of teaching in inner London comprehensives, she went to work as one of the Education Officers at the Centre for Alternative Technology at Machynlleth in Wales in 1989. There she works with school groups on day visits and residential courses, and helps run teachers' courses. She also writes educational materials, does endless guided tours, and says that it is the most interesting job she could possibly find.

Martin Downie trained as a Social Science teacher in Liverpool, qualifying in 1977. He has also studied in Sweden, working there as a potter and as a teacher for a number of years. Upon his return, Martin re-trained as a Design and Technology teacher. He worked for Cheshire Education Authority from 1987 until 1991, and now lectures at Liverpool Polytechnic.

Joan Samuel has a background in Art Education, and has taught in Hawaii and Jamaica (teacher education). In 1988/89 she was Professional Development Tutor in Bedford, focusing on multi-cultural education in predominantly white areas. Joan wrote reports on this for use in Bedfordshire. She is now Deputy Head of Bedford Multicultural Educational Resources Centre and is involved with the Eastern Arts Board.

Helen Abji has a background in English Language Support Teaching and has taught in Tanzania. She is now Deputy Head of Luton Multicultural Educational Resources Centre, working mainly in the secondary field. She recently co-produced a multi-lingual video called 'Safety in the Design Technology Workshop'.

Michèle Young (PGCE Cambridge) has taught in the United Kingdom and the United States. Since 1985 she has been the Artistic Director of Passe-Partout (the Theatre in Education Company) whose vigorous and participatory productions are well known to schools and colleges throughout Britain and in France, Belgium, Kenya, Ghana, Zimbabwe and Japan. In 1987 Michèle was asked by IT to devise and direct the Institution of Mechanical Engineers' Leonardo da Vinci Lecture Series: 'Design for Need', and, following on from that hugely successful enterprise, produced the teaching video 'Is the Price Right?', about small-scale technologies in Kenya.

Mike Watts is currently Reader in Science Education at the Roehampton Institute in London. He taught in inner-city comprehensive schools in London and Jamaica before undertaking a research fellowship at the University of Surrey, where his doctoral work concerned children's understanding of concepts in physics. Through his role as Project Leader with the Secondary Science Curriculum Review he maintained a lively research interest in many aspects of learning in science, in qualitative research methodologies and curriculum development. He has published widely and is the co-author of books and materials on various aspects of education, including a new book on scientific and technological problem solving.

Alan West started his teaching career as a Chemistry teacher, and in due course became Head of Science at a comprehensive school. He became involved in Industry Education work, stimulated by his classroom work on practical problem solving. Links with the Association of Science Education and British Petroleum led to a part-time secondment to the University of Surrey and to work with the Surrey Science and Technology Regional Organisation. He was appointed development officer, and this eventually led to the formation of the Creativity in Science and Technology (CREST) Award Scheme. He is now the National Director of that scheme.

Julian Stapley studied Art and Design in Birmingham and then was a primary teacher for twelve years. He obtained an Open University degree in Science. He became an Advisory Teacher for Design and Technology, and received an M.Sc. in Science/Technology from Warwick University. He has lectured on DES regional courses, and has been involved extensively in Primary INSET. He is now Teacher Adviser for Primary Design Technology in Warwickshire, and is a member of IT's Panel of Primary Specialist Advisers.

Part I   Theoretical Background

Chapter 1 Can We Make the Future Work?
Catherine Budgett-Meakin

A Parable

Once upon a time there was a class
and the students expressed disapproval of their teacher.
Why should they be concerned with
global interdependency, global problems
and what others of the world were thinking, feeling and doing?
And the teacher said she had a dream in which she
saw one of her students fifty years from today.
The student was angry and said,
'Why did I learn so much detail about the past
and the administration of my country
and so little about the world?'
He was angry because no one told him
that as an adult he would be faced
almost daily with problems of a
global interdependent nature, be they
problems of peace, security, quality
of life, food, inflation, or scarcity
of natural resources.
The angry student found he was the
victim as well as the beneficiary.
'Why was I not warned? Why was
I not better educated? Why
did my teachers not tell me about
the problems and help me understand
I was a member of an interdependent human race?'
With even greater anger the student shouted,
'You helped me extend my hands with incredible machines,
my eyes with telescopes and microscopes,
my ears with telephones, radios, and sonar,
my brain with computers,
but you did not help me extend
my heart, love, concern
to the entire human family.
You, teacher, gave me half a loaf'.

Jon Rye Kinghorn, from 'A Step-by-Step Guide for Conducting a Consensus and Diversity Workshop in Global Education', a Program of the Commission of Schools, North Central Association and the Charles F. Kettering Foundation, reprinted in *Global Teacher, Global Learner* by Graham Pike and David Selby, Hodder & Stoughton, 1988.

Introduction

Make the Future Work is a contribution to the technology debate – it seeks to draw attention to the significance and consequences of the application of technology in its widest sense, and to indicate to teachers what their role might be; it is today's teachers and children who can make the future work if anyone can! The authors bring to the debate a wide range of expertise and knowledge – as a glance at the Contents shows. This introductory chapter aims to highlight certain issues that are explored in the book, and to explain the 'vision' behind it.

'Technology 5–16 in the National Curriculum', as well as its forerunners, have been catalysts for much creative and imaginative thinking about the role of technology in the world – in our immediate environment, as well as in school. As David Hicks says in the handbook for the *Global Futures Project*:

> Technology is the only subject to refer specifically to the need to 'cope with a rapidly changing society' and to 'meet the challenge of the 21st century'. This forward looking emphasis reflects both the contemporary nature of the subject and its concern with problem solving.

The fact that the use of such words as 'artefacts, systems and environments' are constant threads in 'Technology 5–16', and that the curriculum is 'process-led' has led to some radical re-thinking of what technology is all about. For some this has been a difficult and threatening nettle to grasp, and there are many examples of teachers retreating into familiar territory; but others have been excited and stimulated by the potential and range, and are keen to push forward their teaching in response to the intellectual challenge offered. It is probably to those teachers that this book will most readily speak.

Since this book is primarily intended for Technology teachers, it is right to consider what we mean by 'technology'. A useful description comes in Ian Smillie's book, *Mastering the Machine* (1991):

> Technology is a combination of knowledge, techniques and concepts; it is tools, machines, and factories. It is engineering, but it is much more than engineering. It involves organization and processes. It has to do with agriculture, animal husbandry and health. It is often highly resource-specific. It involves people, both as individuals – creators, inventors, entrepreneurs – and as society. The cultural, historical and organizational context in which technology is developed and applied is always a factor in its success or failure. Technology is the science and the art of getting things done through the application of knowledge. It is, according to the American technology historian Edwin Layton, 'a spectrum, with ideas at one end and techniques and things at the other, with design as a middle term'.

The book is divided into two sections: Part I is concerned with the theoretical background and with the range of perspectives that must be addressed, and care has been taken to ensure that the relevance and significance of the topic is clear. Part II is the practical expression of Part I, with case studies, activities and illustrations. Wherever possible and appropriate, information is given in a visually accessible form. Each chapter is prefaced by a short introduction, by way of summary, highlighting similar points in other chapters. Appendices give appropriate resources and useful addresses.

Inevitably – and rightly so, in a book of this kind – there is a wide range of approaches. Some chapters will appeal to some and not to others, but I hope that all will find something of value, stimulus, interest and enjoyment.

I have already referred to the 'vision' of the book. It is this: that we all have to find a way of life that makes it possible for:

- our planet to survive and thrive, as a place able to support human life;
- the inhabitants of that planet to survive and thrive without poverty, injustice and exploitation.

This is the true meaning of a 'global' perspective, and it is closely related to the role and our demands of technology. Some may dismiss this vision as idealistic and naïve, but – and here comes the first (of several) quotations from Fritz Schumacher – 'The real pessimists are those who declare it impossible even to make a start.' If we as educators do not have a vision for the future of the young people in our care, and indeed for young people all over the world, then we are not being faithful to the true vocation of 'education' in its widest sense.

Appropriate technology and its role in sustainable development

Fig. 1.1   What is 'appropriate technology'?

To some the term 'appropriate technology' may be unfamiliar, so this is perhaps the time for definitions. An appropriate technology is one which, in general, meets the criteria shown in Figure 1.1.

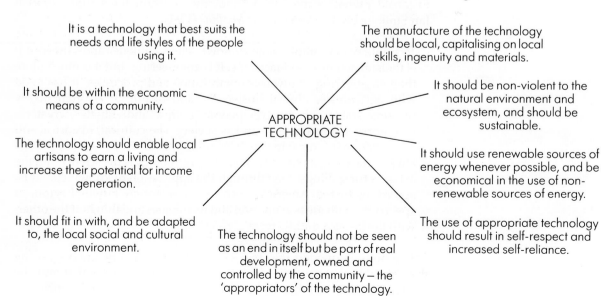

Clearly these 'criteria' are useful tools for looking at technologies in general. How many of the technologies in our industrialised world meet at least some of these yardsticks? How important is it that they should? Which criteria are most appropriate to us?

I would suggest that it would be constructive for us to work towards meeting a great many more criteria than, in general, we do – particularly those to do with the environment – in the interests of global sustainability. It may be that there are other criteria which might be added or exchanged. Such an exercise might be a useful starting point for a class discussion.

And what about sustainable development, making the future work? The Brundtland Report, 'Our Common Future' (1987) defines it as 'development that meets the needs of the present without compromising the ability of future generations to meet their own needs'. That 'development' is meant in a global sense, and therefore applies to us in the rich 'north' as much as, if not more, to those in the poor 'south'. The whole question of inequalities and inequities will be introduced briefly later.

Intermediate Technology – the organisation and its founder

E.F. Schumacher has already been mentioned, so a word about the organisation Intermediate Technology (IT) is apposite here. IT was founded by E.F. Schumacher in 1965 and is a UK-registered charity. IT is now an international development agency of some standing which has as its motto, 'Small is beautiful', the title of the bestseller written by Schumacher in 1973. While still having this as a motto, IT has had to adapt to an increasing demand for its services and approach. IT's aim is to develop appropriate technologies and economic structures which empower people at a local level to become more productive and to earn a living which will be sustainable. IT is, of course, not the only agency involved in this kind of work, but it did blaze a trail for the approach which has incorporated in it a philosophy about long-term sustainable development. Appendix 4 gives more detail about the organisation's practical work.

Schumacher himself was a brilliant economist. He was also a man of extraordinary vision, saying things which still have enormous relevance to our future, and how it could work for the planet as well as ourselves. It was he who coined the phrase 'technology with a human face'. He also said 'the main content of politics is economics, and the main content of economics is technology'.

Much of what he said has enduring significance for us now. An example of his insight was when he looked at the way we consume non-renewable resources. He took as his example the United States, which represents only 5.6 per cent of the world's population. The United States consumes 42 per cent of the world's aluminium, 33 per cent of its copper, 44 per cent of its coal, 33 per cent of its petrol and 63 per cent of the world's natural gas. He said:

It is obvious that the world cannot afford the USA. Nor can it afford Western Europe or Japan. In fact we might come to the conclusion that the earth cannot afford the 'modern world'. It requires too much and accomplishes too little. It is too uneconomic. Think of it: one American drawing on resources that would sustain 50 Indians! The earth cannot afford, say, 15% of its inhabitants – the rich who are using all the marvellous achievements of science and technology – to indulge in a crude, materialistic way of life which ravages the earth. The poor don't do much damage; the modest people don't do much damage. Virtually all the damage is done by, say 15%…. the problem passengers on space ship Earth are the first class passengers and no one else.

That was said as long ago as 1972.

In fact, that quotation indicates the direction of my thesis for this chapter. Over the last century there have been massive technological advances, but the beneficiaries have, largely, been in the 'rich' world – or as I prefer to call it, the 'Minority World'. Use of the term 'Majority World' (which has been used throughout the book) to describe the 70 per cent of

E.F. Schumacher

the world's population who live in what is commonly called the Third World, is accurate in numerical terms and in land-mass terms. And it is useful to realise that we in the 'rich' world are indeed in a minority.

Colonialism – benefit or cost?

The Majority World lives under a legacy of exploitation, which, in many instances, dates back to colonial times. The whole economic system of the ex-colonial African countries was, and in many instances still is, tied to the demands of the 'mother' country. One might be tempted to question the concept of the 'mother country' – what 'mother' would so systematically exploit her 'child'? Raw materials – minerals, food crops (such as coffee, tea or sugar), cotton and so on – were required for the UK or French industrial and technological revolutions. These requirements have not stopped, despite independence. Of course, it would be quite wrong to condemn colonialism completely, but its role in the current world situation must not be underestimated.

Another feature of the world economic system is debt. The massive rise in oil revenues in the 1970s meant that banks had to lend money to earn interest for the oil-rich countries. This money was loaned to Majority World countries, with the ultimate effect of deepening poverty. Indeed there is, in spite of international aid, a net outflow of money to the Minority World, in the form of interest repayments. This has forced governments in the Majority World to take over ever larger areas of land to grow export crops – cash crops – in order to obtain foreign exchange, at the expense of the villagers, whose ability to grow food for themselves is thereby diminished. As John Huckle says in the *Teachers' Handbook* for the World Wide Fund for Nature (WWF) series, *What we Consume* (1988):

Fig. 1. 2 The Majority World

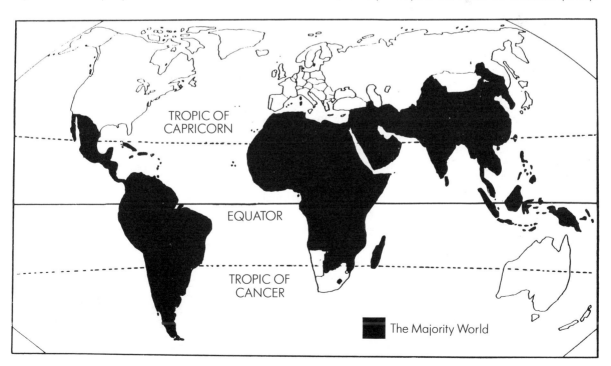

TROPIC OF CAPRICORN

EQUATOR

TROPIC OF CANCER

The Majority World

Over the last four hundred years, the world's people and resources have been progressivly integrated into a single network of economic production and exchange with a related division of labour. Today, commodity chains criss-cross the globe, linking producers of raw materials with processors, manufacturers and consumers. Consumption in one society depends on production in another far away, and consumers generally remain unaware of the social and environmental impacts of what they consume. The world economy is one based on profit and the accumulation of capital; a process which can only take place through unequal exchange and the exploitation of people and environments throughout the world.

Another form of exploitation is that of the natural environment: it is on the natural environment that every one of us, all over the world, ultimately depend. The economic system provides the setting for environmental violence. Short-term profit and short-term gain have been the goals of an economic system which produces entrepreneurs who have ravaged natural resources, without regard for the local inhabitants. As Vandana Shiva says in *Staying Alive: Women, Ecology and Development* (1988):

> With the destruction of forests, water and land, we are losing our life-support systems. This destruction is taking place in the name of 'development' and progress,* but there must be something seriously wrong with a concept of progress that threatens survival itself...the violence...is also associated with violence to women who depend on nature for drawing sustenance for themselves, their families, their societies.

This whole area deserves a far more detailed examination than can be given here. Appendix 2 points to the excellent resources for teachers and pupils, available from development agencies and some publishers – this is what is known as Development Education. But it is important to draw attention to the very complex issues which lie behind the negative images of the Majority World that fill our television screens night after night. It also prompts us to ask 'What is development?' I particularly like the definition I saw in the office of one of IT's partner organisations in Trivandrum, South India: 'Where women, nature and men matter'. The conventional view is that the purpose of development is 'to become like us in the Minority World', but it is increasingly clear that if all the world were to consume and pollute as we do, then the planet's ability to support human life would be seriously in jeopardy.

As Sean McDonagh says in his book *To Care for the Earth*: 'The lifestyle and consumption patterns of many people in the First World are way beyond what the earth can support and can only be maintained by enslaving the vast majority of the world's population.' Or as Gandhi said: 'the earth has enough for everyone's need, but not for everyone's greed.' So far I have quoted visionary remarks from two men now dead (Gandhi and Schumacher), but there is a vast body of confirmatory comment from more recent times: Ted Trainer in *Abandon Affluence* (1985) says:

* See the quotation from Schumacher on page 54.

The characteristic way of life of people in the developed countries involves very high per capita rates of resource use. Typically, these are about fifteen times those of people in Third World countries.... Hundreds of millions of people in desperate need must go without the materials and energy that could improve their conditions while these resources flow into developed countries, often to produce frivolous luxuries.

He updates Schumacher's consumption figures by saying that every year each American uses 617 times as much energy as the average for each Ethiopian. (See the statistics in Chapter 7.)

Needs and wants

This brings me to a discussion of the distinction between wants and needs. This is explored in various ways throughout the book. We all, as human beings, have the same basic needs for survival – clean water, food, shelter and clothing are the basic group, followed by health care, education, community, culture and the opportunity to earn one's living. Transport, energy and access to political influence follow on from those, and there are of course others. Generally speaking, without forgetting members of our society who do live in poverty, we in the Minority World have all our needs met. In addition, we as human beings have legitimate 'wants' – the desire to make the best of our potential, and that of young people in our care, to explore our world, and indeed to be able to enjoy life. What is so glaringly wrong and unjust is that we take these needs and wants for granted, and consider they are inalienable rights, while forgetting the many millions in other parts of the world who cannot assume that such needs and wants will be met. In the Minority World some of the highest-paid people spend their time creating new 'wants'. It is this want-creation industry, more than any other, that stands in the way of making a future that works.

However, it must not be forgotten for one minute that all over the world people have sophisticated cultural and religious structures which are great sources of wealth, and of which we, in our materialism, have largely lost sight. Success for many of us is seen in monetary terms, and in material goods that are symbols of that success. Let me pre-empt David Layton's use of the Gandhi anecdote on civilisation (see page 38). Much of what goes on in our cities is very uncivilised, and it is clear that family norms and values have broken down in many instances across the so-called 'civilised' or 'developed' world. But to those in the Majority World the image portrayed is of glamorous consumption, leisure and abundant material goods which should make us 'happy'. Like our mistaken images of the Majority World, we are represented in too glossy a light which does not reflect reality.

Proactive or reactive?

So what are we going to do about it? Clearly it would be immoral to say to the Majority World, 'Do as we say, not as we do'. Charles Kingsley's Mrs

Doasyouwouldbedoneby and Mrs Bedonebyasyoudid in *The Water Babies* provides a literary analogy of 'Do as I say, because that will be best for us'. Quite rightly India, Kenya and Peru, to take just three examples, want to have access to the technologies and goods which we take for granted. But we must start asking ourselves some very difficult questions:

- Is our way of life sustainable?
- Do we want to transfer it to others if it is not?
- Is there something we should learn from the Majority World that would make it sustainable?
- Do we believe that we should do something about a world economic system that is so blatantly skewed in our favour?
- Do we have a moral right and obligation to examine unjust systems of which we are a part?
- Which technologies are appropriate for a sustainable planet?
- Do we care enough about our planet, and its future inhabitants, to make a real and radical change in our patterns of resource consumption, our consumption of non-renewable sources of energy?
- What role do we have, as educators, for making the future work?

As David Pearce says in *Blueprint for a Green Economy* (1989), in a section called 'Rights of Future Generations':

What is the justification for ensuring that the next generation has at least as much wealth – man-made and natural – as this one? ... what is being said is that we can meet our obligation to be fair to the next generation by leaving them an inheritance no less than we inherited. Moreover, so long as each single generation does this, no single generation has to worry about generations far into the future. Each generation 'looks after' the one that follows.

This brings me back to the brief of this book. As stated earlier, there is great scope within Technology education to examine some of these issues, and indeed there is an explicit requirement to examine technologies from 'other cultures'. Clearly there are dangers inherent in examining a context which is, for the most part, so unfamiliar. But unfortunately it also carries with it the danger, from the development education point of view (with 'problem solving' so much at the heart of much Technology education), that teachers will be under the misapprehension that the whole question of tackling technologies from 'other cultures' is to do with 'solving Third World problems'. Of course there are 'problems' in the Majority World, but many of these are to do with the unfair economic system referred to above, and few are fundamentally of a technological nature. Moreover, it can be argued that the most intractable technological problems are those of the Minority World.

IT education

Intermediate Technology has an Education Office which is constantly publishing school resources (see Appendix 1). These always refer to the locally developed artefacts, systems and environments which are, in many cases, far superior to those imported from the Minority World, because they take account of local cultural and economic patterns. Indeed they are truly appropriate technologies.

The project packs that are being published have great cross-curricular potential. IT, in its work overseas, works in a multi-disciplinary way, with social scientists working alongside people with a technical specialism. This translates into cross-curricular work in an educational context. It is also true that the packs contribute to multi-cultural education, because, wherever possible, there is encouragement to draw on the cultural wealth of a multi-cultural classroom. (But 'all white' schools should not feel that these packs are 'not for them': multi-culturalism is for all.) In addition, because all the packs are 'people-focused', they seem to appeal greatly to girls – which means that there is potential for tackling gender issues in Technology education. Chapter 9 has two case studies which describe what can actually happen in a school (incidentally an 'all white' school) and as a result of extensive in-service development. Chapter 5 contributes an extremely original perspective to the issue of bringing about real change in schools, change that is 'owned and controlled' by teachers.

It is also worth drawing attention to the cross-curricular themes of the National Curriculum. The two that are of particular relevance are 'Economic and Industrial Understanding', referred to in Chapter 4, and 'Environmental Education' – as it says in the *Introduction to Curriculum Guidance*, No. 7 (NCC):

> Faced with air pollution, global warming, the destruction of the rainforests, the advance of deserts, the daily extinction of a species, we have belatedly become aware of how fragile our planet is. Decisions made about the environment affect the quality of life both now and in the future. As consumers and producers, our use and abuse of the environment has wide ranging effects on other people and other living things.... young people ... are [the environment's] custodians, and will be responsible for the world in which, in turn, their children grow up.

So what about the whole question of appropriate technologies in our 'context'? Are the criteria useful tools for examining any technology? How important do we consider the environmental implications of technologies? Environment and development are two sides of the same coin. As David Pearce says: 'One of the central themes of environmental economics, and central to sustainable development thinking also, is the need to place proper values on the services provided by natural environments.' Often one hears that, when comparing the cost of, for example, solar photovotaic energy, and conventional (non-renewable) energy, the latter is cheaper. But what has not been put into the equation is the environmental cost of generating energy from non-renewables. Chapter 4 explores this issue further.

Global warming

We have barely mentioned global warming – surely one of the greatest threats to our planet and its inhabitants. It is generally accepted by a significant body of scientists that global warming will take place unless we make far-reaching changes in our life style. Such changes would include reducing drastically our dependence on the private motor car, taking into account the transport and environmental costs of goods and people moving in large numbers from one part of the world to another – and reducing that flow dramatically. The result would be that we would not be able to buy out-of-season vegetables from far-flung places, and a much more 'local' way of life would become the norm. We would also need to consider our consumption of energy. Far greater use of renewables would be necessary, and desirable, and energy conservation would be automatically integrated into every aspect of our living. I have had a school pupil say to me 'Do you mean we've got to suffer?' Well, no. I do not think, in terms of happiness and fulfilment, I would mind doing without many of the goods and facilities that at present I take for granted. Some things I would find difficult – for example, not using my washing machine so casually, but I am sure I could adapt. There are so many things that could be done without great 'suffering', which would significantly reduce our energy consumption, and therefore our emission of greenhouse gases. Brainstorming with young people about what they could do without is a useful way of introducing these issues, and provokes lively, and perhaps heated, discussions!

If our public transport system received the investment it should, and if it became the norm to use it almost always, and if it were safe to use a bicycle in towns, and they were easily and freely transportable on trains, then the whole pattern of, at least, urban transport would be entirely different, and probably more efficient, comfortable and healthy.

What has this got to do with education and teachers of Technology? I should have thought, a great deal. Educating for a sustainable style of living offers all sorts of exciting challenges and possibilities to the teacher. All school projects would have agreed parameters, with criteria for sustainable soundness being the chief ones. I have already referred to environmental education: Chapter 11 refers to the fruitful link between Creativity in Science and Technology (CREST) and WWF, and to some of the projects which make the concept 'Think globally, act locally' come alive. Likewise Chapters 7 and 8 have a wealth of 'school-friendly' ideas.

Can we therefore make the future work?

If we are talking about making changes in life style, how do we encourage people to make those changes? It does of course need political will, but in a democracy, that comes (or should come), from a change in the ethos and value system of the society. As Jonathon Porritt says in *Seeing Green: The Politics of Ecology Explained* (1984), what we need is 'a different world view'. At the moment economic growth is the guiding principle (see Chapter 4). As Schumacher said,

> The idea of unlimited economic growth, more and more until everybody is saturated with wealth, needs to be seriously questioned on at least two counts: the availability of basic resources and, alternatively or additionally, the capacity of the environment to cope with the degree of interference implied.

The question is: does the economic system direct affairs, or is it a mirror of the predominant ethos? This is the same sort of question that the sociologist tackles when considering the role of education in society – does it mirror society, or can it be an agent for change (see Chapter 2)? Probably the same answer to both questions applies – it's a mix. Certainly schools, on their own, would not be able to promote a sustainable style of living, nor change the economic system, but such changes in social norms and values can be reinforced in school, and sometimes initiate debate among parents. Teachers also are part of the wider society, with their own pressure groups.

Much of this chapter has been concerned with our perceptions in the Minority World. It is therefore appropriate to hear from someone in the Majority World. This is a shortened version of a statement from Bernard Guri, an agriculturalist from Ghana, which first appeared in the Catholic Fund for Overseas Development's (CAFOD) development education magazine, *Link*, in the autumn of 1989, and it provides us with a valuable perspective from the Majority World:

> When I first went to a supermarket in Europe they gave me two plastic carrier bags to carry home my groceries. I folded them carefully and put them under my bed to take the next time I went. But when I did they gave me two more. It was only then that I realised I was supposed to throw them away. But I couldn't bear to…. by the time I came to go home at the end of the year I had a whole pile of bags under my bed to take with me.
>
> This is the way you live… a throwaway life. In the Third World we use something until it is destroyed. We use old cans to carry water; when they get a hole we patch it. You create waste. You make rubbish and pollute the environment.
>
> One day I went to a rubbish tip here in England; I saw all kinds of things which at home I could have fixed and used to furnish my house. Many of them were not even broken. They were just out of fashion. You have created advertisements to tempt yourselves to buy things you don't need. Then you throw away good things to make room for them. You give subsidies to grow crops; then you destroy mountains of them, while in Africa we cannot get enough to eat. You talk about the environment. You talk about the ozone layer and the greenhouse effect. You talk after you have had your fill of chocolate biscuits.
>
> I tell you: there is only one way to solve the threat to the environment. Poverty must be eliminated. How? You must have less. We must have more. You must not give of your surplus. You must sacrifice to give. You must not give out of pity or guilt. You must give out of love. We need your help. But we want to be treated like fellow children of God, not animals on whom you dump food.

This presents us with a real challenge. We are forced to examine our attitudes to 'aid', 'charity', wastefulness and consumption.

Conclusion

Finally, at the risk of repeating myself, we must ask ourselves once more these crucial questions:

- Are we prepared to take seriously the threat to the planet's ability to support its inhabitants, and are we prepared to do something about it?
- Are we prepared to tackle poverty and injustice, not with charity, but by educating the decision makers of tomorrow and by behaving in an exemplary way ourselves?
- Are we prepared to tackle the 'hidden enemies' – apathy, inertia and denial? When doing something, anything, seems so pointless and insignificant, it feels much easier to be an ostrich.

As Edmund Burke said, 'nobody made a greater mistake than he who did nothing because he could only do a little'. The belief that we are powerless is probably the hardest and most insidious obstacle in the path of ensuring justice for all. But it should also present to us, particularly to us as educators, the most stimulating challenge of all – that of becoming empowered ourselves, and of empowering others to ensure that we and our descendants have a planet fit to live in, and that we all, everywhere in the world, have a future that works.

This chapter opened with an educational parable: in the quotation that follows we are reminded of others with wisdom, who understood interdependence. As Chief Seattle has been attributed with saying in 1854 to those in Washington who wished to buy his land:

Teach your children what we have taught our children, that the earth is our mother. Whatever befalls the earth, befalls the sons of the earth, if men spit upon the ground, they spit upon themselves.

This we know. The earth does not belong to man; man belongs to the earth. This we know. All things are connected like the blood which unites one family. All things are connected.

References

Hicks, David (1991) *Exploring Alternative Futures: A Teacher's Interim Guide*.
Huckle, John (1988) *Teachers' Handbook* for the series *What We Consume*, WWF/Richmond Publishing.
McDonagh, Sean (1986) *To Care for the Earth*, Geoffrey Chapman.
'Our Common Future' (The Brundtland Report) (1987) World Commission on Environment and Development.
Pearce, David (1989) *Blueprint for a Green Economy*, Earthscan Publications.
Porritt, Jonathon (1984) *Seeing Green: The Politics of Ecology Explained*, Oxford: Blackwell.
Schumacher, Fritz (1979) *Good Work*, Abacus.
Shiva, Vandana (1988) *Staying Alive: Women, Ecology and Development*, London: Zed Books.
Smillie, Ian (1991) *Mastering the Machine*, IT Publications.
Trainer, Ted (1985) *Abandon Affluence!*, London: Zed Books.

Chapter 2 Beyond the Looking Glass: Technological Myths in Education

Colin Mulberg

In this chapter Colin Mulberg provides an overview of Technology education. As with many of the chapters, the issue of values is raised, and that of responsibility. He picks up the point made in the first chapter, of education as a mirror of society, but replaces education with 'technology', and he explains the range and scope of technology. He draws attention to the issue of the meeting of 'needs' by technological means, and to the business dimension of technology – this links with the economic perspective explored in Chapter 4. The 'design dimension' of technology is alluded to – it is expanded in Chapter 6. But perhaps Colin's most important contribution is to focus on technology in society. He introduces the topic of examining technologies from other cultures, with good examples of how we, in the Minority World, can learn from the Majority World. The range of contexts suggested in the Technology National Curriculum opens a world of opportunity to the teacher with imagination and curiosity.

Despite Technology in schools having such a high profile in recent years, its place in the school curriculum is not a complete innovation. 'Technology' in one form or another has been taught in English state schools since 1902. At that time 'Manual work', such as woodwork, was a required subject for boys, and 'Housewifery' a subject for girls. Though thankfully the gender divisions have been abolished, it is not simply the inclusion of the subject in the National Curriculum that has merited the attention it has received. More importantly, it is the thinking behind it that has brought Technology into the limelight.

The inclusion of Technology within the increasingly crowded curriculum has had to be justified, as it must now compete with other subjects for curriculum time and resources. This is especially the case as Technology is more expensive than other National Curriculum subjects such as Maths and English. The traditional reason that it is intrinsically 'good' for children to learn how to 'make things' is no longer sufficient. In this age of business efficiency and Local Management of Schools (LMS), each subject must fight to defend its territory.

Part of the reason why Technology has secured its place is that the battle for survival has become public. The arguments and counter-arguments have moved far beyond mere discussion of educational principles, to include the value of education in general, the place of education in our society, its value in the marketplace and its contribution to the economy. Indeed, the public's perception of Technology education, which is mostly shared by the teaching profession, is that the subject is so fundamental that it is almost beyond question that it should be taught to children.

Consider the statements in Fig. 2.1 from a recent study of perceptions of Technology in the National Curriculum. Here a range of educationists are giving their reasons why Technology should be taught to school students.

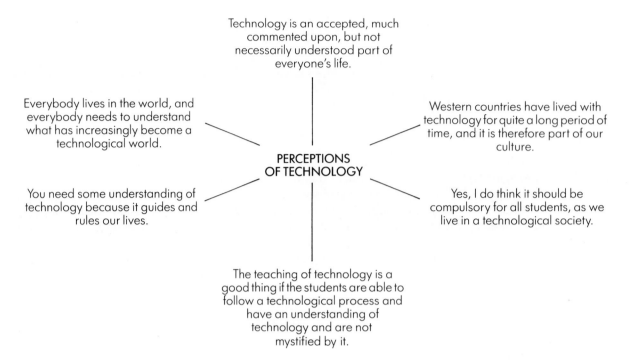

Technology is an accepted, much commented upon, but not necessarily understood part of everyone's life.

Everybody lives in the world, and everybody needs to understand what has increasingly become a technological world.

Western countries have lived with technology for quite a long period of time, and it is therefore part of our culture.

PERCEPTIONS OF TECHNOLOGY

You need some understanding of technology because it guides and rules our lives.

Yes, I do think it should be compulsory for all students, as we live in a technological society.

The teaching of technology is a good thing if the students are able to follow a technological process and have an understanding of technology and are not mystified by it.

Fig. 2.1   Perceptions of technology

These are not specialist statements intended only for discussion within educational circles: they are an extension of public debate. Technology education is perhaps the forerunner of a trend to take education out of the hands of the educationists, and to turn it into public property. Reasons for teaching technology involve an examination of our 'society', our 'world' and what 'rules our lives', and clearly go beyond what is deemed necessary for the education of an individual child. Indeed, as far back as 1976, the 'great debate' in education has linked Technology teaching to the improvement of the country's industry, and ultimately its economic performance. The push towards vocational education through Technology education has grown stronger since then.

Whatever our feelings on the aims of education and the place of vocationalism, the message coming through is clear. Technology in schools is intended to be a *mirror* of technology in the outside world. So as we live in a 'technological society' this must be reflected in a collection of technological experiences in school. It is to be hoped that the lessons learnt from these experiences will then be transferred from school back into society when students finish their education and walk out through the school gates.

Beyond the looking glass

However, most Technology teaching seems to stop short of educating for a technological society by focusing merely on physical technology. This is perhaps the most visible part of technology when we look in the mirror, but it is certainly not the whole picture. Ask people what comes to mind when the word 'technology' is mentioned, and many will say 'computers'. But though to learn about how they work, what they can and cannot do and how to use them in some instances may make life easier, such knowledge is not much help when your telephone bill is wrong, when you have to work the supermarket checkout for the first time, or when the automatic cash machine has just eaten your cashcard.

It is a myth that teaching *about* technology is the same as teaching how to *live in* a technological society. By their very nature societies are complex. Compared with neat, predictable pieces of technology, people are messy and unpredictable. It is a fairly straightforward project to build roads and cars, but to combine them into a successful city traffic scheme that meets the needs of local traffic, commercial traffic and the national network is no mean feat.

If technology is seen in this light, it is certainly an important part of our lives. In fact, it is interwoven into our lives to such an extent, that it makes sense only in the *context* of those lives. A coloured plastic bowl may take on a special meaning only because it is lovingly used to feed the cat; a bicycle can make someone happy as it allows them to ride through the park on the way to work. Without the human connection, the objects of technology are just that – meaningless objects.

But technology can have much more dramatic effects upon people's lives than these simple examples. Get a few insulation calculations wrong and a building may not be technically as good as it should be. This may be enough to turn what should be a comfortable home into a house of misery. Failing to allow for proper maintenance access on tower blocks can affect people on the whole housing estate. Compound these problems with others in many housing estates, and the whole social character of part of a city can be altered.

So, by intermixing technology with people, we get another requirement that cannot be ignored – *responsibility*. It is yet another myth that it is necessary to teach students the technical skills first, and then the moral 'extras' afterwards. It is a sad fact that technology cannot be separated from the responsibility that goes with it. If the mirror of technological teaching is to be a true reflection of our technological society, then the responsibility of the designer must be taught from an early age.

It seems that the only way to abolish the myths of Technology education is to examine in greater depth how technology is incorporated into people's worlds. We need to eradicate the view that technology can be studied separately from the context in which it is used. We need to see beyond the looking glass.

Meeting the people

In common with Alice, as soon as we examine the world beyond the mirror we find it populated by many different kinds of people. But unlike Alice, we do not find a fantasy land populated by fictional characters, but a very real world populated by real people. Each of these people have their own lives, their own histories and their own stories to tell. Truth is certainly stranger than fiction.

The scene is further complicated as each individual has many *needs* that deserve to be met. Some of these are common to all people – food, water, shelter, sanitation, fuel, transport, education, clothing, lighting, communications – and others are more specific; for example, dealing with competing requirements for individual space. People's needs are not just physical; they can be emotional, social, cultural and economic. In people's lives these factors are jumbled together in a variety of ways.

All these needs can be linked to technology, so that a product of technology that fulfils a number of needs can be a joy to use, whereas that which fails to meet people's needs can sometimes be a curse. This is why cycling through the park in the sunshine can bring such an emotional response, as can being stuck in a car in traffic jam.

Sometimes it is possible to get around diverse needs by considering groups of people who all appear to have similar requirements. This can be helpful in certain circumstances, such as considering the needs of disabled people when designing public places. But it must be remembered that such groups are still made up of individuals who may differ socially and culturally, and that the social functioning of groups is itself a complex business. There is also a danger that groups can be over-aggregated to such an extent that their common characteristics become meaningless. This is recognised even within the narrow considerations of retailing, where it is pointless to talk of 'consumers' or 'users', as this is too broad a group; specific consumers would have to be targeted. The design of technology is no different.

If all these different people are considered, each with their own needs, it would be surprising if there actually was complete agreement. Within groups needs will differ, and between groups there is often disagreement. This is the essence of technological design, as nearly every context involves *conflict* over what is required. The greater the number of people involved, the greater the conflict is likely to be. It is possible to get ten people to agree on the route of a railway line. It is harder to get 1 million people to do so.

One of the main tasks facing any designer of technology is somehow to overcome this conflict, and to produce technology that is agreeable to all

Fig. 2.2   'The greater the number of people involved, the greater the conflict is likely to be'

the parties concerned. It is rarely possible to satisfy everyone totally; there is no such thing as perfect technology. But it is normally possible to reach an acceptable *compromise* that goes at least most of the way to meeting people's needs.

A key resource for the designer is therefore *information*. He or she cannot guess what is important in other people's lives, and therefore cannot guess what will improve them, or make them worse. In some cases misinformation does more harm than no information. There is just no substitute for finding out.

It is tempting to use one's own experience in the hope that this represents the experience of others. Often this may be fine, but there is no way of knowing how near or wide of the mark this type of estimation may be. It is a common complaint that those who design ovens are rarely the ones who clean them. When we are dealing with the real world and not a fantasy one, near enough is not good enough.

The obvious way of finding out people's views is to ask them. Yet it is still the exception to involve the end users in the actual design of technology. So, whilst it is now acceptable to involve the community in a community development scheme, it is rare to find prisoners involved in the design of a prison, or patients directly involved in the design of a hospital. And yet these are the very 'consumers' who are the reason for the technology in the first place.

One area of our 'technological society' where finding out is recognised to be of vital importance is that of finance and sales forecasting. It is rare to find investors who would back a product that is uncertain to sell. Accurate marketing is all. If a certain type of shopper requires her or his coffee to stay fresh for longer, then it is sold in vacuum-sealed packs. If this does not keep the product fresh for long enough, then ring-pull cans are introduced. Other parts of the consumer chain may have different needs; if a national supermarket needs coffee to be distributed in twelve-pack trays, then that is what is produced. In this field, to ignore the needs of those you are designing for can mean complete failure. The same is true of all technological design.

It is significant that the strongest move to include people in the design stage has come through the pressure for 'green' products. Supermarkets, oil companies, car manufacturers, toilet-tissue makers and even finance houses are falling over themselves to appear environmentally sound. This may be the decade of enlightenment, but companies from the small through to multi-nationals all know one thing: the market is there. Yet no-one is too sure of what it is that people want. So the right information in the right hands can be as valuable as gold dust.

Politics in the classroom

Nevertheless, whatever information we may find, and no matter how much people may tell us about their lives, we can never capture it all. There is too much detail to take in, order and use. The designer must *distinguish* what is important from what is trivial, what is relevant from what is irrelevant. Only then may he or she proceed to the stage of generating ideas to meet the needs.

'Public transport versus individual health'

This selection process is not automatic. Neither does it happen in a vacuum, but involves notions about what is and what should be. For example, if speed of travel is considered important, then we can find out a vast array of statistics on the time efficiency of different modes of transport. But this tells us little about how enjoyable it is to travel by certain methods. Conversely, if you focus on the type of transport people like the most, this may not be a good guide to explain how commuters actually choose to get to work.

The decisions on what to look for involve *value judgements*, and these values are embedded deep within the technology that is being designed. Whilst the notion of values in technology will be dealt with more fully in the next chapter, the key point to note is that it is another myth that technology in schools is neutral. It is often felt that the 'human' subjects, such as Geography, deal with issues, morals and values, whereas 'pure' subjects, like Science and Technology, are value-free. This is far from the truth.

A look at any Sunday newspaper supplement or television commercial break illustrates that technology involves values. What is often shown is not a pure, rational argument about how a particular piece of technology meets our needs, but images of how the technology could appear to improve our life-style, status and sex appeal. In reality, a car is most likely to be linked to our need for cheap, safe transport, which is flexible enough to allow for the demands of transporting family and goods, and does minimal damage to the environment. Yet advertisements focus on acceleration, luxury, and the fantasy of fast driving through Italian hill

Meeting the needs of all – old, disabled or heavily laden?

villages on the way to the best seats at the opera. In these images, the values have been selected for us; the message is that speed and luxury *should* be most important. If they are not, then we obviously do not deserve to own such a status symbol.

Indeed, it is often a value judgement embedded in technology *not* to treat people as equal. Status can be attributed to technology if the 'haves' can be separated from the 'have nots'. Ownership or control of technology can concentrate power in the hands of the few. This is again the politics of technology, and the technology is designed in such a way for it to be so.

Sometimes the intention may not be so deliberate, but the result is still *technological discrimination*. By only addressing the needs of certain groups of people, the needs of others are ignored. This is again a value judgement. The effect of encouraging some people to use the technology may be to discourage others from doing so. Transport provides yet more examples of this: it has been known for at least two decades that city transport does not address the needs of women, yet little has changed;

buses are not designed for disabled passengers, as many have steep steps; some trains do not have guard's vans and so cannot be used by cyclists; most railway stations are not designed for mothers with prams or pushchairs; and on and on (the list is long).

On a wider front, we can select exactly what needs we wish to address and those we wish to ignore. Likewise, how we address certain needs is also a political issue. Attempts are often made to use technology to tackle social problems, without analysing or drawing attention to what makes them problems in the first place. So, whilst more and more technology is developed, the problem of drink-driving still refuses to go away, and probably cannot be solved by technology. The temptation to go for the technological fix only happens if technology is considered to be separate and outside of society. It could well be that video cameras, body scanners, computerised membership card readers and databases will still not solve the problem of soccer violence.

Educating for a technological society

In the world beyond the looking glass, technology is combined with the physical, social, cultural and economic. Separating out technology loses the *meaning* that it has in a specific time or culture. Technology also involves hard, and sometimes harsh, political decisions that have very real effects on people's lives. We cannot avoid bringing politics into the classroom. If we teach technology, it is already there.

If our mirror is not to distort what we see, then we should teach our students the *skills* they need in order to develop technology in society. They must learn not only how to develop the technology, but also how to combine the technical with the social, the physical with the cultural and the emotional with the economic. They should experience the effect that technology has on people's lives, and they must learn the responsibility that goes with such power.

Good design always involves examples of such integration, but unfortunately it is all too often on a small scale. Vacuum cleaners now account for the need to put the power cable somewhere, and where/how the appliance is likely to be stored. Car dashboards were developed to include a glove compartment that was designed to take all the common European road maps, and a coin tray for parking and toll booths (unfortunately the designers of one well-known French make of car have forgotten that the car is marketed in the United Kingdom, and the coin holder does not hold British coins!). Autobank machines give you your card back first, otherwise you may leave it in the machine after you have taken your money. But this is almost trivial compared to the level of integration that is necessary to make a piece of technology successful.

An effective way of developing these much-needed skills in students is to use the mirror of technology in schools to reflect the level of integration of technology in society. By studying successful and unsuccessful examples, and by designing their own technologies, they will learn not only how technology works, but how it works in a social context. By using technology as a *cultural tool* they will accumulate a wide spectrum of knowledge, and extend their repertoire of successful design strategies.

Technology in other cultures

It is difficult to take a good look at technology in our own society when we are so close to it. It is hard to see what assumptions have been made, what could be challenged, and how things could be different. We need to be able to stand back and look beyond the cultural influences that condition the way we think about technology. This applies especially at school level.

A tried and tested way of getting students to do this is for them to study and design technology for use in other cultures. Right from the start they become aware that technology is used to make judgements about other people; it is easy to assume that a society without 'high' technology is more 'primitive' merely because its level of technology is generally lower than ours. Yet often a lower level of technology may still incorporate a fine balance of needs and available resources. Students soon realise that the assumptions they have made about the needs of other people must be challenged. They cannot guess, and begin to understand the importance of 'mining' for information.

For it is the social ingredient that is the common technological thread that runs through different contexts in different times and cultures. If technology is related to human need, then different technologies across the world can be seen as an attempt to marshal available resources to meet that need. Though we move goods to market in a different way in this country from, say, those in rural India, the underlying aim has common elements. If technological activity is related to other forms of activity, then 'high' technology becomes just *one* way of doing things, and not necessarily *the* way.

Not only is the study of technology in other cultures a way of teaching good technological design principles, but it can also affect the development of technology in our own society. For example, it has taken many years to correct the ignorance of equating 'mud huts' with 'primitive' societies. Yet a study of 'mud' architecture, as still used by some radical modern architects, reveals a wealth of detail about heating and cooling effects, air flow, and passive solar gain, not to mention a greater understanding of the use of communal space. This has helped a new generation of architects to design buildings that absorb the sun's heat in the fabric of the building, and give it out when the sun goes down.

Once technology is related to human activity, it is but a short step for students to see how our lives are connected with lives in other countries. Food is not just grown for our own market and consumption, but is imported from other countries and exported to them. Goods can be produced for a global market, and trade negotiations can attract representatives from many countries. Our technology can have an effect far outside our own land.

Unfortunately it is the ill effects of technology that have brought this to our attention. Our power stations contribute to acid rain, which kills life in the rivers and lakes of others, and we are still learning that radiation respects no international boundaries. Global warming is an issue never far from the agenda of most governments. We still try to export our waste to the Majority World. It is our 'technological society' that has tied our fate to that of others: 'When the rain falls, it won't fall on one man's house.'

Village houses in Zimbabwe
provide useful models for
modern architects

The answer lies not with yet more technology – another technological fix – but with the realisation that our social actions must fit in with those of others. There is a growing call for the recognition that 'global sustainability' is a need for us all, and that this is a social decision. Students must learn that technology on its own cannot bring us any nearer to a solution. *Only a technological society that values our planet can do so.*

Conclusions: the Technology National Curriculum

If students are not to be mystified by technology, then Technology teaching must reveal not just technical detail, but how technology fits into our society. For students to be able to follow a technological process, they must develop the skills to combine the technical with the social, cultural and economic. They must exercise judgement, try to eliminate technological discrimination by resolving conflict through agreed compromise, and show responsibility in the exercise of power. This is the essence of good technology.

The Technology National Curriculum recognises that technology extends into the uncertain realm of people's lives, and that technology has meaning only when people give meaning to it. This is why it places such a large emphasis on students exploring a range of contexts, involving both familiar and unfamiliar situations. More than this, it recognises that it is crucial that students are able to identify people's needs, and those needs involve value, moral and cultural judgements.

The National Curriculum is clear about the importance it places on the skills the student needs to design good technology. It states that before the needs of people can be met, the relevant individuals and social groups must be identified, strategies must be developed for obtaining information from them and involving them in design activity. Then the useful information must be selected before it is applied. Significantly, equal weighting is given to the ability to evaluate the success of technology in meeting needs, and the understanding that different people will have differing needs and values. It expressly mentions the study and design of technology in other times and cultures. If these skills are developed in students, then we will have developed technologists who are ready to go beyond the looking glass into the outside world.

But the Technology National Curriculum does more than this. By focusing on the real world it rejects the fantasy world of some Technology teaching. It demolishes the myths that the technical can be separated from the social, and that by learning about the technical, students will automatically understand its place in society. Out goes the myth that technology in schools is pure and neutral, and does not involve value judgements. Students are expected to examine real contexts involving real people in the real world. It is no longer acceptable to design for imaginary situations, where decisions are arbitrary. It is to be hoped we have seen the last of student projects that claim to help disabled people without the students ever having had real insight into the life of a person with a disability.

The Technology National Curriculum can be taken further. By examining technology in other countries, it becomes even easier to turn the mirror of technology on ourselves. The cross-curricular possibilities make it easier for students to examine their own society. They can learn not just about other people's needs and values, but can start to work out some of their own. They can begin the process of establishing where *they* fit into our technological society. This seems to be a good basis for education.

Selected further reading

Chant, Colin (ed.) (1989) *Science, Technology and Everyday Life 1870–1950*, Open University Press.

Forty, Adrian (1986) *Objects of Desire: Design and Society 1750–1980*, Thames & Hudson.

Papenek, Victor (1984) *Design for the Real World*, Thames & Hudson.

Shepard, Tristram (ed.) (1990) *Ways Forward: Teaching Design and Technology in the National Curriculum*, Stanley Thornes.

Toft, Peter (1987) *CDT for GCSE*, Heinemann Education.

Chapter 3 Values in Design and Technology
David Layton

The power of David Layton's argument for the crucial role and place of values in Design and Technology education is inescapable. He describes how, in the past, values have played little part in technology, and makes a strong case, with the advantage of excellent examples, for their integration into all aspects of technology. He uses the term 'invisible' to indicate how values have been neglected. Though he does not make reference explicitly to the criteria for an appropriate technology, his 'spectacles' to make values visible reveal aspects which are analogous to appropriate technology – for example, his section on 'technology adoption' might have a parallel with the appropriate technology criteria in Fig. 1.1 (page 14): 'suits the needs and life styles of the people using it', and on 'technology obsolescence' with the need for an appropriate technology to be environmentally sound. He develops various categories for different values, and contends that there is a need to 'bring values up into the light of day'.

The centrality of values

Values and value judgements are 'the engine' of design and technology. Judgements about what is possible and worthwhile initiate activity; judgements about how intentions are to be realised shape the activity; and judgements about the efficacy and effects of the product influence the next steps to take. Value judgements, reflecting people's beliefs, concerns and preferences, are ubiquitous in design and technology activity.

The kinds of values called into play are wide-ranging. Some, such as 'the right materials for the job', are technical. Others, such as 'thrifty use of resources', are economic. But the range is much greater than this, including aesthetic, social, environmental, moral and even spiritual and religious values (see Table 3.1).

Table 3.1   *Some different kinds of values in design and technology*

Values	Examples	Values	Examples
Technical	Right materials for the job Improved performance of an artefact 'Neat' solution	Social	Equality of the sexes Regard for the disadvantaged and handicapped
Economic	Thrifty use of resources Maximising added value of a product	Environmental	Ecological benignity Sustainable development
Aesthetic	Pleasing to handle Attractive to look at	Moral	Sanctity of life
		Spiritual/ Religious	Commitment to a conception of humans and their relationship to nature

There are many ways of classifying values – soft versus hard, essential needs versus luxury desires, ultimate ideals versus immediate operational requirements – and the examples in Table 3.1 illustrate that the categories are not always clear-cut. Thus, some environmental values might equally well be classified as social or moral. It is also clear that values are not necessarily mutually supportive, but can conflict. Preservation of wilderness trails may not be compatible with the autonomy of mountain bikers, and incorporation of improved safety standards in the specification for a product may make more difficult the achievement of economic goals. The resolution of conflicting value positions is at the heart of much design and technology activity, and the importance of value considerations is acknowledged in the statutory Order for the National Curriculum foundation subject Technology (see Table 3.2).

Table 3.2   Some statements from the National Curriculum Programmes of Study for Design and Technology Capability

Pupils should be taught to:

Key Stage 1	recognise a variety of forms resulting from people's different values, cultures, beliefs and needs;	Key Stage 3	recognise that economic, moral, social and environmental values can influence design and technological activities;
Key Stage 2	consider the needs and values of individuals and groups from a variety of backgrounds and cultures;	Key Stage 4	recognise the social, moral and environmental effects of technology; recognise potential conflicts between the needs of individuals and society, and negotiate with people having different points of view.

Looking for values in artefacts, systems and environments

An advertisement for a range of office desks in 1961. The desks create the illusion of equality whilst preserving occupational hierarchies

Although value judgements of various kinds drive and determine design and technology activity, when we look at the resulting products we do not see values standing out. Components, structures, materials, configurations and performances may all be directly accessible, but values are only occasionally visible – as, for example, when confronted by different styles of office desks (see photo below). It is rarely obvious what

There's a Status desk for non-stop directors,

. . . for dedicated young executives

and for pretty typists

battles have been fought over conflicting values and how priorities have been assigned to particular values and by whom.

That the values are there, however, is uncontestable, as cross-cultural perspectives frequently remind us. About to board his plane for India at the end of his first visit to the United Kingdom, it is said that Gandhi was asked by an enthusiastic young reporter, 'What do you think about Western civilisation?' After a moment for reflection, the reply came, 'Yes, I think it would be a good idea.' On a lighter note, a cross-cultural review of design and technology responses to the need for mousetraps is equally revealing. There is, incidentally, no recession in the mousetrap industry because of the knock-on effects of two other technologies, one failing (our collapsing Victorian sewers), and one expanding (the take-away food industry). These, with a touch of global warming, have enhanced the domestic demands for mousetraps considerably. Some fifty or more different designs for traps are known, including a French version modelled on the guillotine, an Egyptian one which entombs the hapless mouse in a pyramidal structure, an American model which electrocutes its victims as they run across a metal plate and Britain's most popular 'Little Nipper'. There is an interesting historical dimension to the designs also; medieval traps are resplendent in trip wires, levers and pivoted platforms, whilst an eighteenth-century model called the Deadfall has a 2lb block of solid oak which crashes down on the victim.

Fig. 3.1   Examples of mousetraps

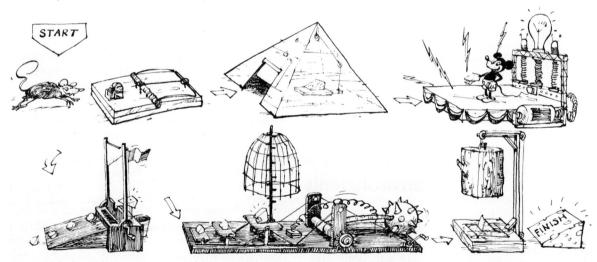

The retention of a distinct product identity with national cultural characteristics, at a time when the markets being served are becoming more global is, of course, a design problem of wide generality. The distinguished Japanese designer Kenji Ekuan, talking about this dilemma, has described Japanese design as being characterised by 'complex simplicity'. He argues that the products tend to be small yet precision-built, lightweight yet robust and energy-frugal, and miniaturised yet of high quality, reliability and functionality, all characteristics which reflect Japanese culture and values (Bethel, 1989).

Making values visible: uncovering values embedded in artefacts, systems and environments

In the course of the national evaluation of the TVEI (Technical and Vocational Education Initiative) curriculum in England and Wales in the late 1980s, a large number of Technology lessons were observed in secondary schools and colleges. Few gave prominence to value considerations (Layton, Medway and Yeomans, 1989). Perhaps because of the low salience and covert nature of values in the products of design and technology activity, explicit reference to them has not hitherto featured strongly in the teaching and learning of design and technology.

In order to 'see' the values embedded in artefacts, systems and environments we need to view these, as it were, through special spectacles. Four ways of looking at design and technology products which help to bring values into prominence are now described.

1   Technology adoption

We are concerned here with the considerations which determine whether or not a technology – a machine, a new crop, a technique, a complex system – is taken up and becomes widely used. The 'success' of a technology is by no means a straightforward matter related to internal technical features alone. More basically it has to do with value judgements of the adequacy and appropriateness of the technology by relevant social groups.

Two examples illustrate this point. Much low-cost agricultural equipment that has been developed is best suited for ownership by contractors or farm cooperatives rather than by individual farmers. An example is that of the forced convection (fan-driven) solar dryer for grain and other crops. Because of the greater productivity of this dryer, compared to traditional methods, an individual farmer will use it for

Because it is only used for part of the year, joint ownership or hire is the solution for some equipment, like this solar dryer being used to dry parboiled potato (papa seca), Peru

limited periods only in the agricultural year. Given the capital outlay involved, joint ownership or hire is an obvious solution. However, despite the effectiveness of the device, adoption is by no means uniform across countries. In some cases, such as Thailand, there appears to be less enthusiasm for joint ownership than in others, like the Philippines, and adoption is less widespread. The value placed on individualism as opposed to collectivism would seem to have a significant influence on the adoption of the technology (Francis and Mansell, 1988, p.82).

The second example concerns personal computers. In the past, some technologies have become dominant by achieving compelling symbolic status within a culture, matching and sustaining, even enhancing, that culture's image of itself. It has been suggested that the widespread use of personal computers, and their transition from a symbol of anti-establishment revolution to one of corporation-defined conformity, is an illustration of this process (Pfaffenberger, 1988).

It appears that a technology becomes a successful working one, in the sense of achieving widespread adoption, when the values embedded in the design are congruent with those of social groups in that particular culture.

More and more cars are being fitted with catalytic converters

Owning a car shouldn't mean disowning the world.

So from September, you can order a new Volvo fitted with a catalytic converter at no extra cost.

Depending on the engine variant, catalytic converters can reduce toxic exhaust emissions by up to 95%.

They don't hinder performance and the cars run exclusively on unleaded petrol.

We've been developing 'cats' in Sweden for over 20 years and we're delighted that the wider availability of lead-free petrol makes them a practical option in the U.K.

We suspect your children will be pleased, too. **VOLVO**

VOLVO ANNOUNCE FREE CATALYTIC CONVERTERS FOR ALL THOSE PEOPLE WHO DON'T OWN VOLVOS.

2   Technology obsolescence or senility

Adoption and obsolescence (or senility) are two sides of the same technology coin. Technology obsolescence refers to the stage in the life cycle of a technology when it becomes dysfunctional with its cultural context. In other words, the values embedded in the technology are no longer congruent with the dominant contextual values.

A glance at the history of technology provides reminders of many devices and manufacturing processes which have been abandoned as no longer appropriate to a changed cultural context. In our own time, the green movement with its concerns for environmental quality is a good example of the way in which new values are bringing about changes in automobile technology, with engines now running on lead-free petrol and exhaust systems fitted with catalytic converters. The same value shift is causing change in other manufacturing industries where 'old' technologies are being adapted or replaced to reduce emissions of sulphur dioxide and the so-called 'greenhouse gases'.

3   Technology transfer

It will be clear from the above that transfer of a technology from the cultural context of its origin and adoption to another cultural context where different values may prevail can bring into sharp focus the values embedded in the design. This is especially the case when transferring a technology from the industrialised Minority World to the predominantly rural Majority World. The usual effect of such transfers, apart from rejection of the transplant, is that the recipient context has to be substantially re-shaped and/or the technology itself has to be extensively re-worked.

An example of the first response is provided by the introduction of the snowmobile into Lapland (see Figure 3.2). The snowmobile, ridden like a motor-cycle on skis, came to prominence in the 1970s especially in the context of winter sports in North America. Of course, as Arnold Pacey

Fig. 3.2   Snowmobiles in Lapland – a mixed blessing for local people!

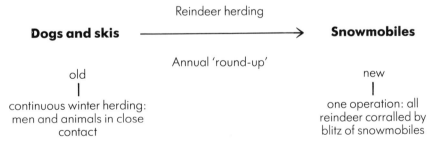

Reindeer herding

Dogs and skis ⟶ **Snowmobiles**

Annual 'round-up'

old   new

continuous winter herding: men and animals in close contact   one operation: all reindeer corralled by blitz of snowmobiles

Consequences

Physiological strain on pregnant female reindeer: fertility and population decline.

Fewer families able to participate in snowmobile herding: cash outlay; expensive maintenance.

Families 'sell out': leads to work as 'waged labourers' or unemployment.

'Successful' herders adopt 'modern' life styles – washing machines, telephones, chain saws, etc.

Egalitarian society ⟶ **Inegalitarian hierarchical society**

(1983, p.3) has noted, a vehicle designed for leisure trips on organised trails between well-equipped tourist centres presents a completely different set of servicing problems when used for heavier work in more remote areas. When they were purchased for use in reindeer herding in Lapland, owners found they needed to carry ample supplies of fuel and spare parts, and to acquire new skills so as to accomplish emergency repairs when breakdowns occurred. The social impact of the snowmobile was, however, much more profound than this.

The capital outlay and the expenses of maintenance meant that relatively few families were able to participate in herding by snowmobiles. Those who adopted the technology found it more economical to work with larger herds; as a result, small farmers, previously with their own herds, were bought out, becoming waged labourers or unemployed. The net effect was that a predominantly egalitarian society, where all owned and worked their own farms, was transformed into a dual society, inegalitarian and hierarchical. Indigenous industries associated with previous methods of herding, involving sledges, skis and dogs, were adversely affected by the change, and an increase in dependency on foreign sources of snowmobiles followed.

The second example concerns Unicef's rural water-supply and sanitation programme in India (see Table 3.3). This involved drilling boreholes and fitting handpumps in order to provide each village with a reliable and health-promoting supply of potable water. However, by the

Table 3.3   Rural water supply and sanitation programme in India

1	Boreholes	– drilling technology – only as reliable as handpumps working the borehole – only as health-promoting a water source as the hygiene practices of users
2	By 1974	– c. 75% of handpumps were out of action in 9000 villages – no system of spare parts provision to village handymen – no training in maintenance – $\$6 \times 10^6$ investment = 9000 holes in the ground!
3	Problem	– pumps were cast-iron copies of European/American farm house pumps – OK for single household but no use when entire village community using them

4   Solutions

Jalna pump	– made by self-taught Indian mechanic – all steel: single pivot handle – connecting rods kept in alignment
Sholapur pump	– better pivot design – manufactured on jigs, therefore uniformity
India Mark II pump	– easily mass produced under Indian conditions – easily maintained – no breakdown guaranteed for 1 year – cost less than $200

– Indian Standards Institute issued full specification for manufacture
– not patented, but quality control by approved suppliers
– installation manual written
– training to staff in installation and maintenance procedures.

Fig. 3.3 The India Mark II handpump

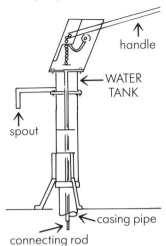

Sri Lankan children in the Anuradhapura area fetch water from an India Mark II handpump, designed especially for heavy use in villages.

mid-1970s some 75 per cent of the handpumps were out of action, and it appeared that an investment of $6 million had yielded little more than 9,000 holes in the ground. One major problem was that the pumps were cast-iron copies of European and North American farm pumps, perfectly adequate for single households, but not suitable for constant operation by members of an entire village. Furthermore, no system for provision of spare parts to the village handymen had been set up, and no training in the maintenance of pumps was offered (Black, 1990).

Local modifications to the design of the pumps were undertaken, some by self-taught Indian mechanics. Eventually these led to a standardised model, the India Mark II pump, constructed to precise design criteria. This was easily mass-produced under Indian conditions, simply maintained with no breakdowns, guaranteed for a year and cost less than $200. The device was not patented, but quality control over shoddy imitations was achieved by publishing a list of approved suppliers. An installation manual was produced and training for staff in installation and maintenance was provided. By 1987, 200,000 pumps per year were being manufactured by forty three registered companies, with exports to countries all over the world.

A further stage in the development of the system entailed the village community taking full responsibility for the operation and maintenance of the pump. Repairs to the India Mark II version required expensive tools and the withdrawal of the below-ground pump parts (for example, to

replace the washers on the piston), a difficult operation requiring lifting gear. Simplicity of maintenance and repair were seen as key factors in the design of a pump for which the village could assume full responsibility. Accordingly, further design changes were made to enable the piston to be withdrawn and a cylinder washer replaced without the need first to dismantle the rising main.

This progressive adaptation of the pump to suit the particular circumstances of Indian villages illustrates two points of importance. First, in order to encourage a long-term, sustainable development perspective on the provision of water in the interests of Unicef's goal of better health for women and children, it was important for the pump to match local resources of materials, energy and human skills. Only then was it likely that the pump would be judged appropriate by those using it. Second, the boundaries of the technology were not drawn around the mechanical features of the borehole and pump. The human, logistical and technical infrastructure to ensure standardisation and quality control of parts, training in installation and maintenance and ready availability of spares was an integral part of the technology. The development of these organisational aspects in accordance with local value preferences was an important contributory factor to successful implementation.

4   Technology and gender

A fourth, though by no means final, perspective on design and technology which helps to make values visible is that of gender. Many of the precursors of Design and Technology, such as CDT and Home Economics, have been afflicted by severe gender stereotyping, although this has most obviously been in relation to materials and skills, rather than values. Wood and metal were associated with the activities of boys; textiles and food materials, with those of girls.

Given the centrality of values in Design and Technology, a question now arises as to whether there are gender differences in the ways in which boys and girls make value judgements. What follows is limited to the field of moral values only. The evidence here suggests that there are important differences which have implications for the teaching and assessment of Design and Technology.

In a remarkable study of the way in which women and men make judgements and act in situations of moral conflict and choice, Carol Gilligan (1982, p.22) provides empirical evidence that 'women (and adolescent girls) bring to the life cycle a different point of view (from men and adolescent boys) and order human experience in terms of different priorities'. She goes on to argue that 'just as the conventions that shape women's moral judgement differ from those that apply to men, so also women's definition of the moral domain diverges from that derived from studies of men' (p.73).

Summarising Gilligan's detailed account, it appears that men tend to construe moral problems in terms of the competing rights of separate individuals; moral development in men is associated with increasing autonomy and with concepts of 'justice' and 'fairness'. In contrast, women tend to construe moral problems in terms of conflicting

Table 3.4   Technology and gender

Carol Gilligan *(In a Different Voice)*:
- studies the way in which men and women develop conceptions of self and make judgements and act in situations of moral conflict and choice
- shows that women bring to life experiences a different point of view (from men) and order their lives in terms of different priorities:

obligations towards individuals with whom they feel connected (see Table 3.4). The problem is one of care and responsibility in relationships, and moral development is associated with the strengthening of empathetic and affectional ties.

The difference is exemplified by the way men and women structure relationships around an idea such as a border between two countries, such as the Berlin wall. Bureaucratic procedures to ensure fairness of access, the nature of conditions for crossing, and the kinds of information to be collected and processed were prominent in the men's reactions. The concerns of the women centred on human relationships, the effects of separation of family members by the border and ways of alleviating distress. The men's world was one of hierarchies; that of women, a world of webs in which goodness was equated with caring.

These moral orientations are undoubtedly reinforced throughout life by the social roles which men and women customarily play. Recognition of the different ways in which the development of moral thinking takes place in adolescent girls and boys has direct relevance to Design and Technology education. At the very least, it encourages a critical view of technologies from the perspective of gender. Although technologies – especially biomedical and reproductive ones (such as *in-vitro* fertilisation, surrogacy, embryo research, the pill) – have major impacts on women's lives, their development has rarely been influenced by female value judgements. As a contributor to a recent collection of essays on women and technology expressed it, 'Women have never lived without technology. Yet we have barely a toehold in the discourse and direction of it' (Hynes, 1991).

Values and models of design and technology activity

It is instructive to review familiar models of design and technology activity from the standpoint of values. Simple linear models (Figure 3.4) imply that 'valuing' is located in the final stage of a sequential process,

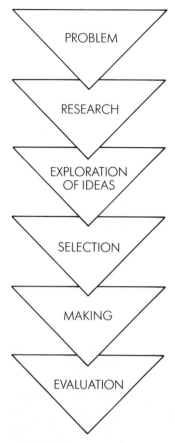

which it clearly is not. Cyclical models (Figure 3.5), whilst relating evaluation to the initial intentions of the activity, likewise do not explicitly acknowledge the all-pervasive nature of value judgements. An interactive version of the cyclical model (Figure 3.6) moves further in this direction, whilst the well-known Project Technology model (Figure 3.7) incorporates 'restraints on technology' under headings, some of which are clearly value-related; for example, financial restraints, personal and social restraints.

All these, and similar models, possess virtues. However, they also display weaknesses in failing to represent some essential aspects of design and technology activity. For example, there is nothing in them to suggest that the activity is rarely, if ever, a solitary and individualistic one, and that team work and cooperation are the norm. Apart from the Project Technology model, there is little to indicate that the activity takes place under a range of constraints which can be conflicting. Most importantly from a values perspective, all are silent on what might be called the politics of the activity; that is, the influences that shape decisions at various points in particular ways.

All technologies carry within them the imprints of the cultures that gave rise to them and what have been described as 'the scars of conflicts, compromises and particular social solutions reached by the particular society' (Goonatilake, 1984, p.121). It is now all too clear that many technologies in the past have not been shaped by long-term concerns about the quality of the natural environment or by what the report of the World Commission on Environment and Development (the Brundtland Report, 1987) termed sustainable development which 'meets the needs of the present without compromising the ability of future generations to meet their own needs'. It is important to recognise here that it could have been, and can be, different. As David Noble (1984, p.xiii), in a fascinating social history of industrial automation, has put it, 'Because of its very

Fig. 3.4   Simple linear model of design and technology activity

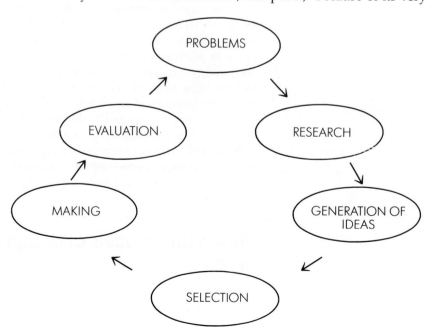

Fig. 3.5   Cyclical model of design and technology

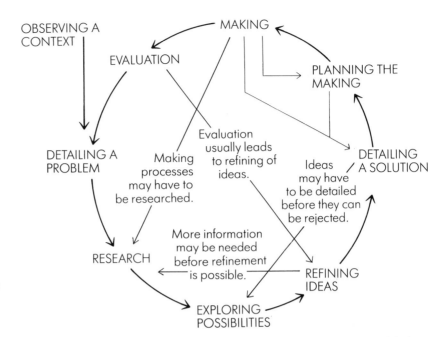

Fig. 3.6 Interactive cyclical model of design and technology

concreteness, people tend to confront technology as an irreducible brute fact, a given, a first cause, rather than as hardened history, frozen fragments of human and social endeavour'. There is nothing inevitable about the form which a technology takes; it is shaped by the value decisions of those in control.

This suggests the need for models of design and technology activity different from those reviewed above. Fortunately there are alternatives available, although they have not so far featured prominently in the

Fig. 3.7 Project Technology model of design and technology activity

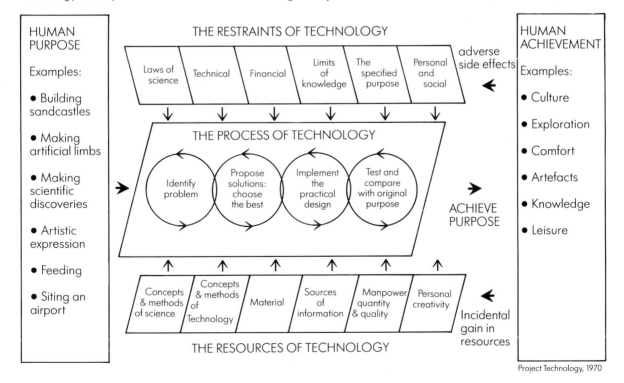

Project Technology, 1970

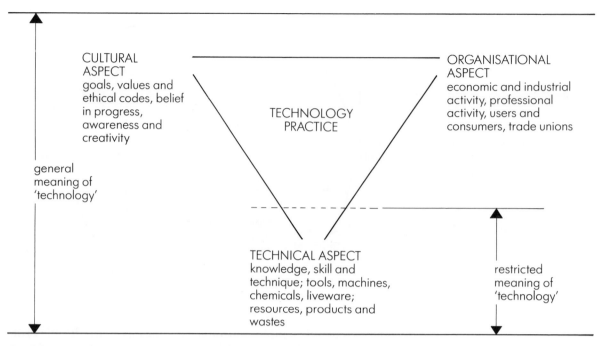

CULTURAL
ASPECT
goals, values and
ethical codes, belief
in progress,
awareness and
creativity

TECHNOLOGY
PRACTICE

ORGANISATIONAL
ASPECT
economic and industrial
activity, professional
activity, users and
consumers, trade unions

general
meaning of
'technology'

TECHNICAL ASPECT
knowledge, skill and
technique; tools, machines,
chemicals, liveware;
resources, products and
wastes

restricted
meaning of
'technology'

Fig. 3.8   A model of design and technology activity which gives prominence to values (after Pacey, 1983, p.6)

teaching of Design and Technology in schools. One model which attempts to incorporate the values dimension of design and technology practice is shown in Figure 3.8 (Pacey, 1983, p.6).

Although not a model in the sense of the diagrammatic ones above, the attainment targets and statements of attainment for the design and technology capability profile component of the National Curriculum foundation subject Technology also embody value considerations in relation to both the processes of undertaking design and technology activity and the eventual product. The most obvious example is Attainment Target 4, which states that 'Pupils should be able to develop, communicate and act upon an evaluation of the processes, products and effects of their design and technological activities and of those of others, including those from other times and cultures'. This opens the way for critical reflection not only on all aspects of pupils' own work but also on the value options and decision processes which have empowered technological developments in the past and which are doing so today.

Value judgements are not confined to AT4, however, and the statements of attainment for other attainment targets contain many references to the role of values in all aspects of design and technology activity. Table 3.5 provides examples related to AT1 which requires pupils to 'be able to identify and state clearly needs and opportunities for design and technological activities through investigation of the contexts of home, school, recreation, community, business and industry'.

Resolving value conflicts

There are at least two ways in which value conflicts can arise in design and technology activity.

First, the values embedded in an artefact, system or environment may conflict with those of the cultural context into which it is introduced. In

Table 3.5 Statements of attainment from Attainment Target 1 which refer to values

Level	Pupils should be able to:	Level	Pupils should be able to:
4	(d) explain that a range of criteria which are sometimes conflicting must be used to make judgements about what is worth doing;	7	(a) analyse information of several kinds and draw conclusions about the needs and opportunities for design and technological activity, recognising and resolving conflicting considerations about what is worth doing;
5	(b) recognise that economic, social, environmental and technological considerations and the preferences of users are important in developing opportunities;	10	(c) make reasoned judgements about what is a subject for design and technological activities and what is better dealt with in other ways.
6	(b) explain how different cultures have influenced design and technology, both in the needs met and the opportunities identified;		

GCSE Criteria (Draft)

Aims specific to Design and Technology Syllabus

Design and Technology 'is unavoidably concerned with identifying and reconciling conflicting human values'.

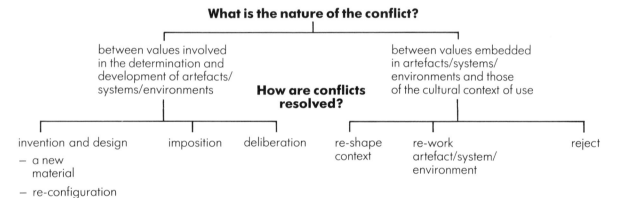

Fig. 3.9 Resolving value conflicts

exploring the issue of technology transfer we have already seen that such conflict can be resolved by (a) re-shaping the context and/or (b) re-working the artefact, system or environment. Alternatively, the artefact, system or environment may be rejected.

A different and more familiar case occurs when values conflict in the development of an artefact, system or environment. An example here might be the design of a new type of passenger seat for large, intercontinental aircraft. Considerations would include ergonomics (the seat must enable passengers both to sit upright and to recline in it); safety (the design must reduce the risk of injury if a crash occurs and also minimise fire hazards); aesthetics (fabrics used must be not only fire-resistant but also available in attractive colours and patterns and comfortable to sit on); and economics (the cost of the seat must not be exorbitant, and its shape and size must permit a large number of passengers to be carried). The list could easily be extended by other requirements, such as the adaptability of the seat to the needs of the elderly and the physically handicapped. However, it is already clear that the potential for value conflict is considerable in a situation such as this.

As the GCSE Draft Criteria remind us under aims specific to Design and Technology syllabuses, all design and technology 'is unavoidably concerned with identifying and reconciling conflicting human values'.

It has to be said that there is no simple and certain way of achieving this latter goal. Indeed, it represents a major intellectual challenge of design and technology activity. Reviewing ways in which value conflicts have been resolved in the history of design and technology, three broad strategies can be discerned.

1   Invention and design

Technological solutions to problems are sometimes possible as when a new material is developed which combines sought-after properties previously unavailable in a single product: for example, a fabric which is inexpensive, hard-wearing, can be woven into attractive patterns, takes coloured dyes well and is fire-resistant.

An example from a different field is the development of oral rehydration therapy (see Figure 3.10). This avoids the need for intravenous treatment and hence skilled administration, so bringing a cheap, effective and easy-to-use technique to those in countries where diarrhoeal diseases previously affected nearly 500 million children annually and resulted in some 5 million deaths from dehydration and loss of electrolytes from bodily fluids.

Numerous examples involving semiconductor electronics demonstrate how technological advances have made possible the achievement of ever more complex and rapid communication goals. Similarly, developments in food technology and the invention of the microwave oven have made compatible the enjoyment of attractive, home-prepared meals and a lifestyle for men and women which is both occupationally and recreationally full.

It is, perhaps, misleading to separate design from invention; the two are intimately related. But sometimes no new physical principle or technological device is needed to resolve a value conflict, merely a reconsideration of design. The case of Duracell torches illustrates the point. According to Professor Nick Butler, head of the design consultancy which Duracell approached, the starting point was an economic concern to sell more batteries. This led to an investigation of products using batteries and particularly of torches and their functions. The need to angle the beam of a free-standing torch in order to illuminate an object, so

Fig. 3.10   Oral rehydration salts – the life saver. Unicef shows people how to make their own solution

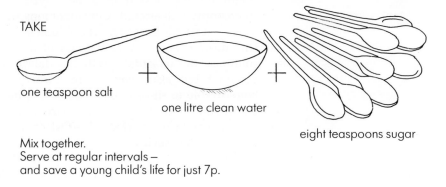

TAKE

one teaspoon salt   +   one litre clean water   +   eight teaspoons sugar

Mix together.
Serve at regular intervals –
and save a young child's life for just 7p.

leaving both hands free to work on this (for example, changing a wheel or a fuse in the dark), led to a completely different design for a torch, increasing its usefulness and, simultaneously, the sale of batteries and the product range of Duracell (Butler, 1988).

2 Imposition

Especially when different values are being espoused by different individuals or social groups, value conflict may result in winners and losers. The history of technology provides many examples of the resolution of value conflict by the exercise of power, one group imposing its preferences on others. David Noble's (1984) study of industrial automation identifies competing versions of the technology, favoured by managers and shop-floor workers respectively, the latter losing out to the former. The transfer of some technologies from industrialised nations to countries in the Majority World, a process which the recipients have on occasions termed 'technology dumping', may also be seen as a form of imposition.

3 Deliberation

The explicit recognition of different value positions and the exploration through communication – explaining, persuading, criticising, accepting criticism, negotiating – of ways of achieving compromise and consensus is the third means of dealing with value conflict. There is, of course, no guarantee of consensus, especially when the task is to achieve a balance between values associated with shorter-term, immediacy, survival perspectives and those related to longer-term, future need perspectives. Economic and moral or environmental concerns may pull in opposite directions with no obvious quick resolution possible. What might emerge from such value confrontations is, at least, the re-prioritising of values in the positions of individuals and groups and also the reorientation of research efforts to achieve technological and design solutions to the conflict.

Teaching 'values'

Accepting what has gone before about the centrality of values and the inescapable need to resolve value conflicts in design and technology activity, there remains the question of how to teach 'values'. This is contested territory, where marked national differences occur (Cummings, Gopinathan and Tomoda, 1988).

Recognised positions include the following.

1 Values clarification

This approach assumes that the act of valuing involves

- choosing freely from alternatives after thoughtful consideration of the consequences of each alternative;
- prizing, cherishing, being happy with the choice, willing to affirm the choice publicly;

- acting, doing something with the choice, repeatedly, in some pattern of life (Raths, Harmin and Simon, 1976, p.78).

Teaching involves asking pupils questions which 'clarify' their responses; for example, 'Did you consider any alternatives?' The approach avoids moralising, criticising, giving values and evaluating; it places the responsibility on the pupil to decide what she or he wants. It has been criticised for ethical relativism and for treating all values as having the same status. 'Something clarified, but nothing valued' is the judgement of one opponent of the approach. It also does not offer a satisfactory way forward for resolving interpersonal value conflicts.

2   The cognitive-developmental approach

This is based on Lawrence Kohlberg's theory of moral development, which identifies three levels:

- a pre-conventional level where moral decisions are based on self-centred needs and are egotistic;
- a conventional level where moral decisions are relativistic, being characterised by conformity and maintenance of 'law and order';
- a post-conventional level of principled morality, decisions being based on values shared with others, leading to universal ethical principles.

Teaching involves confronting pupils with 'moral dilemmas', which stimulate the development of moral reasoning to the next stage.

The theory is concerned with moral values only and hence relates to a limited part only of the values territory over which design and technology activity takes its participants. Even within the realm of moral values it has been criticised for lack of universality; much of the empirical work on which the theory is based was undertaken with adolescent boys, the majority from the United States. Not only have people like Carol Gilligan raised doubts about the theory's applicability to girls, but also its validity in cultural contexts very different from the United States, such as China, has been questioned (Meyer, 1988, p.118).

3   Indoctrination

Both values clarification and the cognitive-developmental approach are opposed to the third position: inculcation and indoctrination. Teaching here is by exemplary models, exhortations to emulation, rewards and punishments.

Before dismissing the approach as educationally inappropriate in a democratic society, however, it is worth reflecting that some of the values associated with design and technology activity are non-controversial and, from the standpoint of the health and safety of pupils, leave little room for personal clarification or cognitive development. Cleanliness and hygiene in handling food materials and tidiness as a contributory factor to safety in workshops are values which are perhaps best inculcated at an early age.

There is a clue here to the solution of the general problem of teaching values in design and technology. There is no 'one right method', and the range of different kinds of values encountered suggests that a variety of approaches is needed. Inculcation may be all very well for 'safety' but would be inappropriate for 'economic efficiency' in conflict with 'ecological benignity'. At the end of the day it will be a matter of professional judgement when clarification of a pupil's declared value position is merited and when confrontation by a staged 'moral dilemma' might assist development.

The final message is that there is a need to bring values up into the light of day in the teaching and learning of design and technology; to categorise them (for example, moral, economic, technical and so on) and make them the subject of deliberation and critical reflection between pupils and pupils and between pupils and teachers.

References

Bethel, D. (1989) 'From Heart to the Market: Different Senses of Design between East and West', *The Times Higher Educational Supplement*, 17 March 1989, p.15.

Black, Maggie (1990) *From Handpumps to Health: The Evolution of Water and Sanitation Programmes in Bangladesh, India and Nigeria*, New York: United Nations Children's Fund.

Butler, Nick (1988) 'Design and Technology: The Industrial and Economic Context, in *Design and Technology in the National Curriculum*. Report of the National Conference held on 29 November 1988. London: The Standing Conference on Schools' Science and Technology, pp.2–11.

Cummings, W.K., Gopinathan, S. and Tomoda, Y. (eds) (1988) *The Revival of Values Education in Asia and the West*, Oxford: Pergamon Press.

Francis, A.J. and Mansell, D.S. (1988) *Appropriate Engineering Technology for Developing Countries*, Research Publications Pty Ltd, 12 Terra Cotta Drive, Blackburn, Victoria, Australia 3130.

Gilligan, Carol (1982) *In a Different Voice: Psychological Theory and Women's Development*, Cambridge, MA and London: Harvard University Press.

Goonatilake, S. (1984) *Aborted Discovery: Science and Creativity in the Third World*, London: Zed Books.

Hynes, H. Patricia (ed.) (1991) *Reconstructing Babylon: Essays on Women and Technology*, London: Earthscan Publications.

Layton, D., Medway, P. and Yeomans, D. (1989) *Technology in TVEI, 14–18, The range of practice*, Sheffield: The Training Agency.

Meyer, J.F. (1988) 'A Subtle and Silent Transformation: Moral Education in Taiwan and the People's Republic of China', in W.K. Cummings, S. Gopinathan and Y. Tomoda (eds) (1988) *The Revival of Values Education in Asia and the West*, Oxford: Pergamon Press, pp.109–130.

Noble, D. (1984) *Forces of Production: A Social History of Industrial Automation*, New York: Alfred A. Knopf Inc.

Pacey, A. (1983) *The Culture of Technology*, Oxford: Basil Blackwell.

Pfaffenberger, B. (1988) 'The Social Meaning of the Personal Computer', *Anthropological Quarterly*, 61: pp.137–47.

Raths, L., Harmin, M. and Simon, S.B. (1976) 'Selection from Values and Teaching', in D.Purpel and K.Ryan (eds) *Moral Education … It Comes with the Territory*, Berkeley: McCutchan Publishing Corporation, pp.75–115.

Further reading

A book which is informative, insightful and highly readable on the subject of technology and values is *The Culture of Technology* by Arnold Pacey (see above). The author has worked in a number of countries overseas and with organisations including Oxfam and the Intermediate Technology Development Group.

Chapter 4 Appropriate Technology? Appropriate Economics!

Ken Webster

Ken Webster's chapter brings Economics and Technology education together. With Economic and Industrial Understanding as one of the cross-curricular themes, it is important to draw into technology its economic aspects, and Ken does this convincingly. There is also a link with the values aspect addressed by David Layton. Schumacher was an economist, and he certainly realised the interrelationships with and interdependence of technology, values and economics. Ken makes the point that appropriate technology will not work without appropriate economics as its framework, and provides us with a stimulating analysis.

Introduction

The whole point is to determine what constitutes progress.
Schumacher, *Small is Beautiful*, p.131

Absolutely. The crisis, as Frijof Capra has recently noted, is 'a crisis of perception'. How we see the world changes our actions towards it. We can't escape the issue. For Capra and Schumacher, of course, technology and economics are at heart a matter of values. That much must be obvious in this book.

Schumacher noted something rather special about the relationship between technology and economics. He said that the main content of politics was economics, *and the main content of economics was technology*....[my emphasis], so if we would reform our technology we must reform our economics. Attempting appropriate technology with inappropriate economics is nonsense. Literally – it is trying to add apples to pears. The reason for this revolves around perceptions. Conventional economics sees itself as largely value-free and objective, and transferable to any issue which involves choices about the uses of resources. Appropriate technology is a technology which recognises and accepts that human needs and values are complex. It does not pretend that what *can* be done *should* be done, or that a technology is necessarily transferable. Appropriate technology is often specifically local, needs-orientated, resource-efficient and flexible. Would that our economics were so humble!

Luckily for the teacher and the pupil 'appropriate economics' – by which we mean the economics that recognises the role of values – has been really acknowledged and incorporated in economics in schools for the last decade. Its practitioners perhaps don't always recognise its strength or its application to technology in general – though this is changing rapidly – or its particular relevance to appropriate technology in the school. But it lives for all that. It is called and sometimes abused by the name 'economic awareness'. Grasp it with both hands and be thankful that you can make it what you need and make it serve the new perceptions of this decade. It is an appropriate idea, small, flexible and empowering. It is consonant enough for most purposes with the 'new economics', the economics-as-if-people-mattered growing out of Schumacher, Illich, Max-Neef, Ekins, Henderson, Anderson and the rest, to be used in education as a synonym.

There is just one proviso: education is not about selling a solution. The ideas and methods used in 'appropriate technology', need to be evaluated in the same way as any other technological option, however much opposing appropriate technology sounds like promoting the inappropriate. The same is true of what might go on under the heading of 'new economics', or in school the more radical elements of 'economic awareness'.

But a short introduction begs a lengthier explanation. And for that we must detour by a few hundred years. What went wrong with economics? And why should I care?

The present — is it a product of the past?

The perception or 'big idea' of Newton and Descartes was that the universe was like a huge machine, set in motion by God and left to unwind. The world had an objective existence, independent of mind. The mind of man contained the spark of the Divine. The highest human faculty was therefore that of the rational, intellectual, scientific. The machine metaphor was very successful. With it came the desire to analyse the parts in order to find the fundamental elements of the machine and explain the mysteries of life, the universe and, it was assumed, 'everything'.

Science, of course, seemed unbounded. Bacon had asked his followers to torture Nature's secrets from her. Nothing, it seemed, was beyond the scientist. And so God and the idea of Divine spark was dispensed with. Experimentation, the testing of ideas in the laboratory and the clear superiority of the whole business for those involved in it actually promised the earth. Some years later this attitude found its fullest expression in the confidence of the Victorian era in 'Progress'. It deserved the capital P. Economics as an emerging 'science' in the nineteenth century was not immune from the central idea that the world was a machine. It was not a human-scale economics which emerged.

In the first place, since economics had to be a science if it was to have credence alongside the hard sciences, it should be value-free – objective. It must deal with that which is measurable and comparable. Therefore it was not the business of economists to suggest what should be done or ought to be done. Rather, they would analyse the situation, discover the mechanism and report what they found.

They assumed that the individual person in the marketplace was rational and sought to maximise her satisfaction. The person was a cog in the economic machine which was the market – the place where the invisible hand of supply and demand directed resources to their most efficient uses. The world was assumed to be full of individuals driven by this same desire to maximise their satisfaction. Consumption would stand for satisfaction and money for a measure of consumption.

Production of those goods and services which are chosen by these individuals is guided by the the market. The producer who can produce more at lower money cost stands to sell more and increase her profits and at the same time increase the overall level of consumer satisfaction. All will be well, we are told, if the machine can run freely, if the market functions without restriction. Competition will see to that. Everything of value will be exchanged in a market, and the measure of that value will be monetary. Work is transformed from the 'good work' that Schumacher saw as a central purpose of human life – that combination of hand and brain in the creative satisfaction of a wide range of human needs – to a soulless, mechanical repetition put at the service of limitless wants. The 'machine age' demanded 'machine people', for after all they were mere factors of production, something to be subverted to the logic of an endless increase in consumption. And this is what is wrong with much of conventional economics: the assumptions are a poor match with the reality of production consumption and exchange, and with the complexity of human life.

Even more seriously, no distinction was made between useful or frivolous goods, between useful creative work and mind numbing work; between a resource and an ecosystem; between consumption and satisfaction; between needs and wants. When Descartes began the process which led to the exclusion of God from the universe, the same world view generated the economics of more and more – because it seemed that goods and services were all that could fill the vacuum left by the end of God. Descartes said, 'I think therefore I am.' By the late twentieth century one maxim being offered was, 'I consume therefore I am.' Meaning is now defined by what is owned. A magazine reports that 95 per cent of teenage women in the United States (in a 1991 survey) list shopping as their most enjoyable activity. This illuminates an economics for the poor in spirit, an economics for those with increasing expectation but declining hope. Even in the former Communist block a satirist in one cartoon has attached a Tesco supermarket bag to the outstretched arm of Lenin. The implication is clear.

There is a phrase describing the pursuit of 'more and more': economic growth. It is defined loosely as the increase in the production or real income of a society over time. It is the generally accepted criteria for a successful economy. Not having as much economic growth as some other country or society is almost sinful. Philip Larkin echoed some of the expectations and consequences of such an economic age in his reflection of post-war England, 'Going, Going':

Lenin with a Tesco supermarket bag on his arm!

> The crowd
> Is young in the M1 café;
> Their kids are screaming for more –
> More houses, more parking allowed,
> More caravan sites, more pay.
> On the Business Page, a score
>
> Of spectacled grins approve
> Some takeover bid that entails
> Five per cent profit (and ten
> Per cent more in the estuaries …

The economic growth drug seems hard to resist until one realises that in the generation of the figures no allowance is made for what goods are produced or what services are provided. The production of garden gnomes is seen as every bit as valuable as that of bread. The services of the men and women paid to clear up a polluted river are counted as contributing to growth, but the help I give my neighbour with his hedge is not, simply because no money changed hands. The figures are not a reliable indicator of quality of life, but then economics as conventionally taught is very bad at dealing with quality. And as long as it is narrow and quantitative, it is as inappropriate in the school as it is in the community.

Fig. 4.1   What is the real cost?

What is the real cost of...? damn something in one easy step. Call it 'uneconomic'. Misuse the language further by identifying the economic with the financial. Yet we all do it from time to time. Installing energy-saving equipment, say solar panels or insulation, may appear unjustifiable if the financial saving is less than the outlay and the interest which would accrue to the money if it had not been spent and kept in a bank. But the equation is narrow in the extreme. What is the real price of the energy used? It is more than the money price.

- The environmental deterioration is not 'costed'.
- The relative scarcity of the fuel in the long term is not 'costed'.
- The efficiency of the conversion of fuel to useful work is not fully reflected in the price.
- It may be that the energy-saving equipment is very economic – in the true sense of the word.
- It may be a very low-cost solution overall to the heating of water in a cool climate.

This problem with costs happens on the large scale as well. It seems strange, if one stands back from it, how anyone can equate the flooding of a large valley with economic growth on the basis of the flow of electricity coming from the hydroelectric plant at the barrier and the use of that power for new industries. The calculation is naïve. Some examples:

- What is the value of the lost land in production?
- What is the value of the timber and other resources on or under the land?
- What is the value of the wasted labour of the displaced people, perhaps once farmers and now dispossessed. What proportion of the new energy supply will reach the new workplace?
- What are those new workplaces producing anyway?
- Who will really benefit from the energy and the new goods produced?
- How long will the reservoir last before it is silted?
- How long will the turbines survive the clogging water-weed?

All these factors have contributed to the debate around dams on large rivers. They are part of the true economic equation. (The journal *The Ecologist* carries numerous examples of the true economic folly of large dams: see for example, Vol. 18, Nos 2/3 (1988), p.56, 'The Proposed Three Gorges Dam in China').

An appropriate economics sets its work in the heart of the economic, environmental and social web, not apart from it. An appropriate economics welcomes debate on the radical alternatives offered by, among others, the appropriate technologist.

If the technology teacher still finds that he is largely in agreement that the aim of economic activity is the pursuit of increased consumption through increased production (without all those other rather awkward questions troubling him) then he may have to stick to conventional business and economic analysis. In this traditional analysis projects are

What is the real cost of a dam like this one, which is part of the Mahaweli Scheme in Sri Lanka?

judged by such indicators as their marketability, by their return, by their profitability or their market share. The measure is money, tempered perhaps by cost/benefit analysis – a guessing game in which the values which cannot be monetarised are translated into money.

It might be useful to reflect on what Schumacher had to say about cost/benefit analysis:

> [cost/benefit analysis] is generally thought to be an enlightened and progressive development….in fact, however, it is a procedure by which the higher is reduced to the level of the lower and the priceless is given a price….to undertake to measure the immeasurable is absurd.
>
> (*Small is Beautiful*, pp.37, 38)

What am I bid for the ozone layer, anyone?

In a way, the rather mundane financial issues – codified in the National Curriculum as the economics and business studies elements in the Technology document – are questions for accountants and perfectly proper within their own sphere, but the financial questions are not the only or even the first questions asked by the appropriate technologist. The financial aspect is just too narrow a focus. Failing to see the broader picture, failing to look at the social, environmental and ethical elements reduces the economic element of a technological exercise to a dangerous and ultimately irrelevant role. The appropriate technology movement puts people first. A human-scale economics is similarily 'self-limiting'; that is, it recognises that infinite growth on a finite planet is impossible, it is quality-driven, people-centred, needs-centred; it promotes self-reliant, community-based development. It recognises cooperation as well as competition and that cooperation extends to the planet: in the immortal phrase 'economics as if people mattered' – and as if the planet mattered. It is also very difficult because the appropriate technologist or

A bicycle is cheap and adaptable, and even more so with a bicycle trailer

economist has to see things another way. She needs a new world view.

Essentially, Schumacher rejected the old but still dominant world view. He did not perceive of the world as a machine. He saw it more as a self-regulating, self-limiting system. An economics for which efficiency means processing world resources for consumption as quickly as possible and at a low financial cost is a nonsense on a planet characterised by such self-regulating systems. An ecologically sound economics describes efficiency in terms of the least use of resources necessary to accomplish the task – coupled with the greatest access to, and control by, the populace. As Illich noted in a slightly different context: *'Democracy moves at the speed of the bicycle.'* A bicycle is cheap in resource use, accessible to most people and under their control, yet capable of being adapted towards a variety of ends. This would obviously not be true of the motor car, which is elitist, expensive in terms of resources, and controlled, some cynics would say, by the industry which supplies them.

So the teacher or student of Technology will wish to apply broader, explicitly value-centred economic criteria when assessing what to do and evaluating how it went. An outline of how this might work in practice is described below.

First there is an opportunity to explore some current definitions of economic awareness, to see if they might serve for our appropriate technology/appropriate economics purposes.

Economic awareness is variously defined, but as Ian Pearce, then of Educating for Economic Awareness (National Curriculum Council), stated, it involves 'a critical awareness of the use of scarce resources' (*Forum*, spring 1989).

In part, 'economic awareness' reflects a desire for the re-interpretation of economics as a discipline. This was the view of Hodkinson and Thomas, editors of the *Economic Awareness Journal* (issue 1). More generally, however, the quotation below summarises one consensus view of what economic awareness attempts:

> Everyone is involved in the economic system which operates within our society. We all face the problems of limited resources and have to take decisions about how best to make use of those resources.
>
> Economic Awareness is about making these issues more explicit and understandable. Extending pupils' levels of economic awareness empowers them to
>
> - more fully understand the economic forces which shape their lives;
> - develop a questioning attitude towards what is best and for whom;
> - make more effective use of the resources currently available to them and to develop positive strategies for improving their own quality of life and that of other people;
> - understand the effect of their own actions on society locally, nationally and internationally.

(Tameside LEA/the Economic Awareness Teacher Training Programme)

The key word here is perhaps 'best'. It begs other questions such as 'best for whom?' 'how do we know?' 'how do we decide?' This definition of economic awareness does not seek to diminish but enhances and does justice to the central notion of value. It should therefore be of use to an appropriate technologist. In a recent booklet prepared by the BBC and the Economic Awareness Teacher Training Programme (ECATT) to accompany an economic awareness in-service video, further clues are given to the potential for the idea in technology by its emphasis on open enquiry methods.

> As a result of Economic Awareness work with teachers and pupils we have come to recognise that because pupils make their own sense of the economic system their Economic Awareness can can only be extended if they are prepared to reassess that interpretation.
>
> Four characteristics of this process appear to be essential. Each is dependent on the others and all give meaning to the view of learning which underpins the teachers' concerns and their professional discussions.
>
> 1 Active involvement by the pupils. The need for the active and interested involvement of pupils in the learning process is now generally accepted. In Economic Awareness it will be secured through their response to information/stimulus/problems/tasks.
> 2 Communication, group work, co-operation, presentation. These skills are a key feature of participatory learning. In Economic Awareness pupils will develop them if they are encouraged at all times to share the sense they have made and the conclusions they have reached about the information/stimulus/problems/tasks.
> 3 Thinking creatively and critically. In Economic Awareness, thinking creatively and critically means more than sharing opinions. It means encouraging pupils to reassess the sense/conclusions they have made

about the information/stimulus/problems/tasks by creating opportunities for them to:
- check the statements they make against evidence
- listen to other people
- carefully clarify the meaning of words, categories, statements
- identify the basis for similarities and differences between their own and other people's opinions, views and statements.

4 Extending Economic Awareness

A pupil's Economic Awareness will only be enhanced if the process of active participation, skill development and thinking carefully and critically result in:
- the formation of new categories and concepts
- the identification of new relationships.

Economic awareness can be an essential part of the technological process, the design 'loop' – the cyclical process beginning with establishing needs and opportunities for technological activity and ending with an evaluation of the technology in action. Economic awareness might be apparent in two respects. First, the teacher might want to use the economic awareness emphasis on a general critical awareness, when tackling AT1, 'establishing needs'. An initial example is from Britain, a country, perhaps unexpectedly to some, in special need of appropriate technology. Let's imagine that a group of young people are examining the pedestrian and other traffic flows in and around the school, the way people move, and how and why. The youngsters are invited to make an assessment of the needs of the users of the space, and suggest possible solutions to the apparent congestion – a not uncommon brief.

Hang on! The teacher will have spotted the assumptions underlying the exercise. Here are two: that there is congestion, that these people need to be here at all. Ridiculous? Let's think about it, taking the last assumption first. Schools are only one solution to the need for 'education'. Illich saw them as part of the problem. Why have your young people and the staff here at all? Would another arrangement be a better use of resources?

Congestion around the school in this sense is a symptom of a problem, not its root. Secondly, is there congestion? What is congestion? Are there advantages to being close together. Does it help a sense of community? Does it enable information to flow more rapidly? Who is really losing out if the place is crowded? Do the teachers and the pupils actually value the break between lessons or delay as they reorientate for the next session in a crowded school? Where exactly is the problem?

These are all questions about alternative ways of organising human activity. They are all driven by considering alternative uses of resources.

They are clearly economic questions. What are the implications?

Allowing pupils to stop the world to ask basic questions such as these is the sign of an economic awareness perspective in full flow. The alternative is the assumption that efficiency in education is directly and causally related to contact time. It might be argued that talking to crowds of people reduces pupil contact time and therefore efficiency, thereby wasting resources. Typically, the socially attractive areas of a city are the 'crowded' markets or narrow streets, or pedestrianised areas of the 'old'

A group of young people have been given a rather closed brief about how to cross a river (below left). Currently a small rope ferry does the job. The suggestion is that a bridge would be one solution with which to cope with increased traffic. A map describes the site. Resources are made available in the form of straws, glue, string, cardboard and so on. Each of these is priced per unit. Not unsurprisingly the young people quickly decide that a bridge is needed or that the teacher seems to want this(!). A design-and-make with an evaluation through testing to destruction is undertaken. Smiles all round?

But what was the broader situation? A more detailed map would make an important contribution (below right). Clearly, most teachers will recognise the inadequacy of the process here. Not enough information has been given in order to establish the problem. This point, the establishment of needs and opportunities, is well made in the Technology document. But the appropriate technologist is armed with other resource questions.

- Was a bridge really a good option?
- Who decided? Why?
- Will it meet local needs?
- Will it be under local control?
- Have local people been consulted?
- Would another solution use fewer resources, energy, and so on?
- Is the cost of a bridge really only that of the materials to build it?

- What would be the state of the road if the bridge encourages rather than relieves traffic?
- What is this traffic anyway? Is it bullock carts and pedestrians or lorries or cars?
- Could another route be devised? And so on.

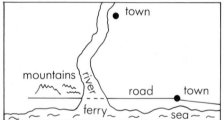

Original map – little information

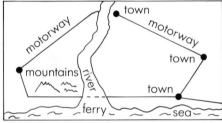

This one has useful extra information. Is a bridge needed?

A richer experience for the young people is surely possible when a full consideration of the social, environmental and financial costs and benefits are debated. This is the role of the new economics: to ask different qualitative as well as quantitative questions. Perhaps it is part of thinking globally, getting a broader, longer-term, more sensitive perspective and applying it locally, to the problem in hand.

Appropriate technology attempts to be context-driven, not driven by what is technologically possible. In the same way an appropriate economics will use but not rely on the techniques of business and narrow notions of efficiency. Paul Ekins, at a conference run by *Beshara Magazine* in 1990, described a possible model for the relationship between economics and other elements.

Each of these elements is connected internally to the others. The value of a project can be 'tested' against the way in which it reflects what Ekins assumes to be a necessary balance between each element. Try taking the original brief for the river crossing and marking it on the model according to how it reflects each component. How will it look? How should it look in an appropriate technology sense? What will you do to ensure more balance?

(after *Gaia Atlas of Planetary Management*, Gaia Books, 1992)

Fig. 4.2   *An example of missing and reclaiming the opportunity to ask appropriate economic questions*

town.... Even an out-of-town shopping mall attempts to be a 'village on a motorway junction'.

Let us move now to another part of the design loop – that of AT4 – evaluating the artefact, system or environment. Imagine we are in 'Nepal' and that a new stove has been designed by the villagers with the aim of saving fuel-wood – which in this village is in short supply. The stove is shown to provide as much heat as before with less wood and more precision, and the smoke from the fire is safely carried out of the kitchen area, thus cutting down the risk of acute respiratory infection. How could we evaluate this design?

Traditional measures suggest criteria such as cost of materials and labour and the alternative use of time thus absorbed. Maintenance costs and the saving of fuel-wood and the time taken to collect it might also be an important criteria. However, there might be other issues. The assumption underlying the use of the stove may be that it was for cooking only. The old, inefficient stove may also have heated the house. The new one doesn't do this as well. Perhaps more firewood will be used to try and lift room temperatures. Perhaps the assumption was that one new stove suited all kitchens. This may have ignored differences in techniques and requirements by different ethnic groups. The Tamang people may cook differently, using different sized pots and so on, than, say, the Sherpa or the Nepali.

Fig. 4.3 For teachers: a summary of the sorts of questions which might inform the approach to the 'resource' or economic dimension of technology

Economic awareness in this general sense allows young people to ask more profound questions as part of their technological work and to make better sense of the world around them – by engaging them in it. Pupils often ask, 'Why are we doing this?', 'What is it for?' Economic awareness as part of an appropriate technology might be part of an appropriate answer to their question, one which they could discover for themselves in cooperation with their teachers and others because it concerns life as it is lived rather than life as it is imagined it is lived.

- Who is involved in the decision making?
- What decisions are they making?
- What resources are being used?
- What are their reasons for making these decisions?
- What is likely to happen as a result of these decisions?
- What private and social costs and benefits will result from these decisions?
- What possible alternative decisions could be made?
- Which is the 'best' decision and why? Best for whom?
- What constraints limit and influence the decisions?
- What, if anything, can be deduced about the values behind the decisions?
- Are there any implications for government action and policy?
- Is there sufficient unbiased information to make an informed evaluation of the decisions?
  (From a model developed by The Northern Ireland Council for Educational Development, NICED)

Further reading

Economic Awareness Journal, Longman (3 times a year).

Ekins, Paul (ed.) (1986) *The Living Economy*, London: Routledge & Kegan Paul.

Illich, Ivan (1974) *Energy and Equity*, London: Calder & Boyar.

Lutz and Lux (1988) *Humanistic Economics*, Bootstrap Press.

Robertson, James (1978) *The Sane Alternatives*, published privately by James Robertson.

    (1990)*Future Wealth*, London: Cassell Educational.

Schumacher, E.F. (1979) *Good Work*, Abacus.

    (1974) *Small is Beautiful*, Abacus.

Webster, Ken (1990) *Energy, Economic Awareness and Environmental Education*, London: World Wide Fund for Nature.

Chapter 5 Managing Change: Appropriate Technology Strategies for Change in Your School

Alan Dyson

This chapter breaks some very interesting and original new ground: Alan Dyson is drawing an analogy between the appropriate technology movement in a Majority World context – appropriate technology for sustainable development – and curriculum innovation. He argues that much curriculum development has been imposed from above, particularly in the recent past, without consultation with those at the grass roots, or chalk face, of education. He looks at some telling examples, both from 'development' and from education, to illustrate his thesis. He proposes an appropriate technology consultancy to enable the lessons from 'development' to be understood, and applied to education – a good example of 'learning from the Majority World' and perhaps 'sustainable curriculum development'. It might be hoped that readers of this book will take up his challenge. The second case study, in Chapter 9, tells the story of curriculum innovation where teachers have been involved all along the line.

Appropriate technology: for some or for all?

Fig. 5.1   Case study: the Bakolori Dam

I want to begin this chapter by looking at two case studies. The first is an example of economic development going somewhat astray in the Majority World.

Bakolori is the name of a dam and an agricultural project in Nigeria. It represents a huge investment in Italian dam-construction technology by the Nigerian government, aimed at giving poor farmers in an arid region a steady supply of irrigation water, and hence increased crops, security and wealth.

Unfortunately, the scheme has had many problems. In the first place, the building of the dam meant the destruction of what was already there. Fourteen thousand displaced persons had to be resettled (in much poorer conditions), and the land of many others expropriated. Farmers participating in the irrigation scheme itself had to suspend their work while land-levelling went ahead, and that process hit many problems and delays, and hence was much slower than expected.

When the scheme was finally completed, the expected benefits failed to materialise. Instead of using irrigated techniques to obtain a second crop, as expected, the farmers persisted with traditional methods, producing only one crop per year. Part of the explanation is that disputes with the river basin authority over subsidies were unresolved. However, a more fundamental reason is that the type of agriculture demanded by the scheme is congruent neither with the farmers' work habits nor their beliefs. The dry season was a positive advantage to them since they could use it for other tasks, such as repairing their houses or going on pilgrimage to Mecca; while the successful use of irrigation required them to work outdoors at night, something of which their religious beliefs made them afraid.

(Adapted from Buxton, 1983)

The second case study concerns a project with which I was recently involved: the adoption of a new course and its associated materials by an English school.

The Salisbury School Humanities Project is a set of teaching materials developed by a 'show-piece' school in the North of England and exported on a trial basis to a series of other schools which had expressed an interest. Salisbury School placed a high premium on collaborative group work, open-ended exploratory activities, and pupil-directed learning. The materials reflected this, and consisted of ideas for teachers and guide-sheets for individual or group explorations rather than the more traditional work-sheet type of resources.

West Hill was one of the participating schools, its senior management team attracted by the idea of being able to bolster the curriculum in an area they perceived to be underdeveloped. Unfortunately, their expectation of being able simply to import a packaged course was largely unfulfilled. The first problem was that the use of the Salisbury Project meant that teachers had to abandon what they were already teaching, much of which they had developed for themselves and viewed with considerable pride. To add to their resentment and frustration, Salisbury School were less efficient than they might have been in the production and delivery of supportive resources, with the result that West Hill teachers were unable to plan effectively.

Even when the resources did arrive, the problems continued. Although they had been developed in Salisbury School for a very similar group of pupils, the West Hill teachers found them difficult to use in their classrooms, claiming that they were unsuitable for 'our children', and therefore engaged in a fairly major re-design of both content and materials. That re-designing resulted in a course which in many ways resembled that which it had replaced. In one respect this was particularly true: the Salisbury School guide-sheets, predicated upon exploratory learning and collaborative group work, were turned into traditional question-and-answer written work-sheets.

One by one, the West Hill teachers abandoned the Salisbury Project until, within less than a year of its introduction, it was no longer being used in the school.

Fig. 5.2   Case study: the Salisbury School Humanities Project

The first of these case studies is quite clearly a classic case of what we have come to understand as inappropriate technological development. It has all the signs and symptoms: a Minority World technology imported into a Majority World country; a failure to consult local people; a conflict between local practice and belief and the new technology; and the ultimate rejection of what was designed to help but has actually harmed. As citizens of a Minority World country ourselves, we are quite comfortable with the notion that appropriate technology actually means technology that is appropriate in the Majority World, a less sophisticated, less efficient technology that is simply a bridging device to lead the Majority World gently towards high technology.

What strikes me, however, and perhaps will also strike the reader, are the similarities between the two case studies, even though they are situated, actually, worlds apart. The leadership of the school was as well-meaning, but as non-consultative, as the leadership of the state; the Salisbury Project was as much an alien technology to the West Hill teachers as the dam was to the Nigerian farmers – and it was no more perfect or totally efficient; the existing practice of the teachers was destroyed as surely as the farmers' land; there was as much conflict between the West Hill and Salisbury views on how children should be taught as there was between the single- and double-crop forms of agriculture; and the 'Salisbury technology' was undermined and rejected in just the same way as the irrigational technology in Nigeria. If, in other words, the Bakolori Project is an example of inappropriate technological development, then so is the Salisbury School Project. The notion of 'appropriateness' is not one which has relevance in the Majority World alone; it applies to the 'sophisticated' institutions of the Minority World also. It applies to us.

Technology, culture and organisation

It is not difficult to see why this should be so. Elsewhere, David Layton has demonstrated how value judgements are inseparable from technological judgements. A technology is not simply a 'way of doing something' that can be assessed solely in terms of its effectiveness and efficiency. It implies judgements about what is worth doing and what costs are worth paying to get it done. Pacey (1983) has a model which I find helpful (see Figure 3.8). He argues that any given form of what he calls 'technology-practice' has three 'aspects':

- its 'cultural aspect' – its implicit goals, values and belief;
- its 'organisational aspect' – the particular forms of social and economic organisation which it demands and within which it is embedded;
- its 'technical aspect' – the skills, techniques and knowledge which tend to spring to mind when we think of technology.

It is fairly evident how this model can help us account for the fate of the Bakolori Project, as a clash between different techniques of agriculture embedded in very different cultures and forms of organisation. However, I would argue that the model also tells us much about what happened to the Salisbury Project. For teaching also is a technology; it comprises a set of techniques, a 'process', for getting certain educational things done, and, inevitably, particular techniques of teaching are embedded in fundamentally different forms of organisation and culture.

The difficulty arises when a form of teaching technology is imported into a school or department which does not share the cultural values or does not have the forms of organisation implied by that technology. Returning to our case study, it is not an adequate explanation to say that the West Hill teachers were simply too incompetent or unintelligent to implement the Salisbury Project. Rather, they did not share the values of that Project. Salisbury School was concerned with teaching as the facilitation of a process of collaborative exploration. Its teachers organised their classrooms and developed the necessary professional skills to enable them to actualise those values. West Hill saw teaching as the transmission of skill and knowledge. The teachers put themselves at the centre of their lessons and were highly skilful information-deliverers. If they did not have the skills or the classroom organisation to facilitate group work and discussion, it is because group work and discussion were inconsistent with their own values.

Appropriate technology and the teaching of Technology

I would like to focus this argument more precisely on the teaching of Technology in the National Curriculum. Perhaps another case study would help:

> Some years ago, I was teaching in a comprehensive school. One of my tasks was to help Anne, a girl with spina bifida, to integrate into mainstream classes. Her disability was relatively mild, but it did mean that she had very limited use of one hand and was slower than most children in manipulative tasks. Most teachers coped with this problem well, and the girl participated fully in CDT, Art, Science, PE and so on. However, the Home Economics teacher found her extremely difficult to teach and invited me to observe her in a cookery lesson. What I saw was most revealing.
>
> The lesson began with a demonstration by the teacher to the class of the techniques that would be used in that day's cooking. Then the teacher wrote the recipe on the blackboard, and set the pupils to cook for themselves. From that point on, the teacher spent her time moving rapidly from pupil to pupil, chivvying, trouble-shooting and occasionally taking over. The lesson was conducted at great pace, with constant reminders about the time and what point in the recipe the pupils should have reached.
>
> Inevitably, Anne fell behind the others almost immediately. The teacher visited her more often than the others and gave her more help. Frequently she took a task over from Anne and completed it for her in order to help her catch up. I suggested, as delicately as I could, that it might be worth letting Anne find her own way of doing things, and that the teacher might leave her to work undisturbed for the rest of the lesson. My colleague looked at me somewhat incredulously, but stayed away from the girl for the next five minutes, casting anxious and increasingly frustrated glances in her direction. Eventually, as Anne fell further and further behind, she could bear it no longer. She came across to her, completed her preparation, and put her cake into the oven only slightly behind the rest. By the time the bell went for the end of the lesson, Anne had a beautiful cake to take home with her. 'There you are,' said the teacher, triumphantly. 'I had to do that or she would never have achieved anything.'

Fig. 5.3   Case study: Anne

I tell this story not to denigrate a colleague, but to raise for the reader what I believe are important questions:

- What values did my colleague have: what did she see 'teaching Home Economics' as being about, and how did she believe children learn?
- How did her organisation of her classrooom and her skills as a teacher reflect those values?
- What values did I display, and how compatible were they with those of my colleague?
- How do you imagine this teacher is coping with National Curriculum Technology?
- If you were her Head of Faculty, what, if anything would you do?

It seems to me that the value clashes that are evident in this case study are particularly relevant, because they are precisely those which have arisen in the implementation of National Curriculum Technology. On one level, the attainment targets and programmes of study simply specify content to be taught and some techniques for teaching it. To that extent, they are a specification of a technology of teaching. However, in common with all technologies, they are imbued with values, beliefs and organisational implications. Implementing National Curriculum Technology, therefore, is not simply a matter of following what has been laid down; rather, it is about individual teachers and whole schools coming to terms with a complex value system and with the organisational arrangements which are necessary to realise those values. There is no reason to believe that this will be any simpler – or, come to that, more successful – than the equivalent undertakings in the Bakolori or Salisbury Projects.

It might, therefore, be useful, to try to clarify what some of the implicit values and organisational assumptions of National Curriculum Technology might be. In the list which follows I have kept in mind what seem to me to be important differences between the National Curriculum approach and the more traditional approach that characterised the teaching of 'practical subjects'. However, readers may find it useful to construct their own lists of those implicit features of National Curriculum Technology which are most dissonant with their own values and those of their colleagues:

- Technology is part of the core of human learning and understanding.
- Technology is essentially process-based.
- Technology is a coherent curriculum area rather than a collection of disparate subjects.
- Assessment in Technology is assessment of the process as well as the product.
- Pupils need to be given a degree of autonomy in order to participate in the process.
- Many of the key skills in Technology are cross-curricular rather than subject-specific.
- Classrooms and workshops need to be organised flexibly in order to realise pupil autonomy.

- The principal teaching strategies will focus on facilitating group work, negotiation and active learning rather than on the didactic transmission of knowledge.
- Technology is an entitlement for all pupils, with a scope and content which is the same for all; therefore the teaching of Technology has to be differentiated.

However much one finds oneself in tune with these implications, the unavoidable reality is that many primary teachers tackling Technology for the first time, and many secondary teachers with a background in one of the traditional 'practical' subjects, are going to find National Curriculum Technology not just technically difficult, but also alien to their views of teaching and learning. If, like my colleague, they see Technology as a collection of unrelated practical skills which have to be transmitted by the teacher to groups of pupils working in the same way and at the same pace, then there is no point of engagement between them and the National Curriculum. This, I would suggest, presents enormous problems for those amongst their colleagues who have to manage the school's Technology programme.

The dimensions of this problem become clearer when we consider not just the content and implications of National Curriculum Technology, but also the process whereby it has been introduced. For what is overwhelmingly evident is that, whatever the merits of the attainment targets and programmes of study – and I believe them to be many – the implementation process itself constitutes a massive exercise in inappropriate technological development as potentially disastrous in its own way as any that we have seen in the Majority World. It is an imposition, ultimately by legal force, of a technology developed in one culture – the culture of politicians, administrators and experts – on an entirely different culture – the culture of schools and teachers. It is, moreover, a means whereby the imposing culture seeks to meet its own needs – for accountability, standardisation and 'quality assurance' – by claiming to identify needs in the recipient culture. And, finally, it is a technology which the imposing culture has handed over to 'trusties' in the recipient culture – namely, senior and middle managers in schools – for detailed implementation, thus placing those 'trusties' in the invidious position of being given power to impose the will of others on their own colleagues.

Put in this way, those managers would seem to have to choose between two alternatives. One is to acquiesce in the process of inappropriate development. At best this is likely to lead to their colleagues' simply 'going through the motions' of the National Curriculum. At worst, it may produce hostility and outright rejection. The other alternative, it seems to me, is to find some strategy for integrating the imperatives of the National Curriculum with the values and organisational characteristics of particular teachers and particular schools; in other words, to transform the National Curriculum into what we have learned to call an 'appropriate technology'. This is an exceedingly difficult and delicate task, and if it is to be successfully carried out, then we must not be afraid to learn the lessons of technological development in what at first sight appear to be very different contexts. It is to these lessons that I now wish to turn.

Appropriate technology consultancy

Emerging experience in the development of appropriate technologies in the Majority World is beginning to yield some sort of consensus as to how non-governmental agencies and other consultants can best facilitate that process. That consensus can, I believe, help us to understand how consultants in schools – and by this I mean senior and middle managers working with colleagues as well as external consultants such as LEA officers and higher education trainers – might develop appropriate technologies of teaching.

There are now a number of consultancy guidelines. Trainer (1989) offers eleven guidelines for appropriate development; Madeley (1991) suggests 'Twelve guidelines for reaching the poorest'; Remenyi (1991), although taking the narrower focus of credit-based income-generation programmes, none the less offers a series of recommendations that seem to have some relevance for schools; and, of course, there are the nine criteria of appropriateness that Intermediate Technology (see Chapter 1, page 14) has distilled from the seminal work of E.F. Schumacher.

Useful as all these guidelines are – and I would certainly recommend readers to pursue them for themselves – I personally prefer a series of 'lessons' learnt by Robert Chambers (1988) from a review of case studies in technological development. Partly this is because they are simple and therefore manageable; partly it is because they arise not from general principles alone, but from detailed field-work. Chambers suggests that five central lessons emerge:

- The first lesson is to follow the learning process rather than the blueprint approach.
- The second lesson is to put people's priorities first.
- The third lesson concerns secure rights and gains for the poor.
- The fourth lesson is to achieve sustainability by starting with self-help.
- The fifth lesson is the importance of calibre, commitment and continuity of [consultancy] staff.

(pp.8–13)

It might be useful to consider each of these 'lessons' in turn and see what we can learn from it about managing development in schools.

1 Follow the learning process rather than the blueprint approach

It is helpful to view this 'lesson' in the light of Pacey's model of technology. Any given form of technology practice may appear relatively straightforward and unproblematical; there may seem to be little to learn about it beyond the immediately apparent techniques which it comprises. However, we know from Pacey that it will be embedded within a cultural and organisational context the full extent of which may remain hidden. It may well be only the implementation of those techniques in a different situation which reveals the full implications of the technology practice as a whole, and which allows the consultant to learn what they might be.

Moreover, that situation itself comprises cultural, organisational and technical characteristics which are certain to be exceedingly complex and largely hidden from view. It is simply not possible, therefore, to predict with any certainty how a particular technology will impact on a particular situation. That is something which must be learnt. It follows that it is fruitless and even dangerous to try to draw up a blueprint for implementation; a learning-process approach is the only one that is likely to work.

What does this mean for the school 'consultant'? Contrary to common belief – a belief which, I would argue, is central to the National Curriculum – development in schools is not simply a matter of devising 'best' solutions to teaching problems at some central point and then instructing teachers in how to implement them locally. The solutions themselves have to be local and have to be arrived at by a process of learning about the local situation. This learning process does not have to be one of detached observation. On the contrary, we now have an extensive literature of action research (see, for instance, Elliott, 1991) which links learning and practice. Although there are many varieties of action research, they are all ultimately founded on a cycle of acting in a real situation, carefully observing the consequences of action, and acting again in the light of what has been learnt (see Figure 5.4).

I would suggest that the 'technical aspect' of action research is not difficult, but that its cultural and organisational aspects may well be. It is, by definition, an ongoing, never-completed and messy process. Consultants in schools, especially managers, have to give themselves permission to be process-orientated, to tolerate ambiguity, to allow things to develop. After all, as technologists, they are now in the business of granting precisely this sort of permission to their pupils.

2   Put people's priorities first

At a time when the educational world is coming to terms with external imposition as the major means of development, it is salutary to consider Chambers' realisation that development actually springs from what people want, not from what others believe they need. The literature both of economic development in the Majority World and educational development in schools is littered with examples of projects which have failed to take account of people's wants, and have ultimately been rejected by precisely those they were supposed to benefit. Our own initial case studies serve as prime examples.

Managers, advisers and in-service trainers are accustomed to saying to teachers, 'This is what you have to do'. Chambers is suggesting a prior question: 'What do you want to do?' Again, this is not a question that is difficult to ask in a technical sense; but it may well be culturally difficult. It requires a significant shift in power relations and a willingness to listen to our 'clients', even though, 'they do not conform to what outsiders are looking for' (Madeley, 1991, p.22). As Chambers puts it: 'too often outsiders prevent themselves learning from the poor because they adopt the role of teacher or disciplinarian' (p.10). How many of us have fallen into precisely this trap in our work with fellow teachers?

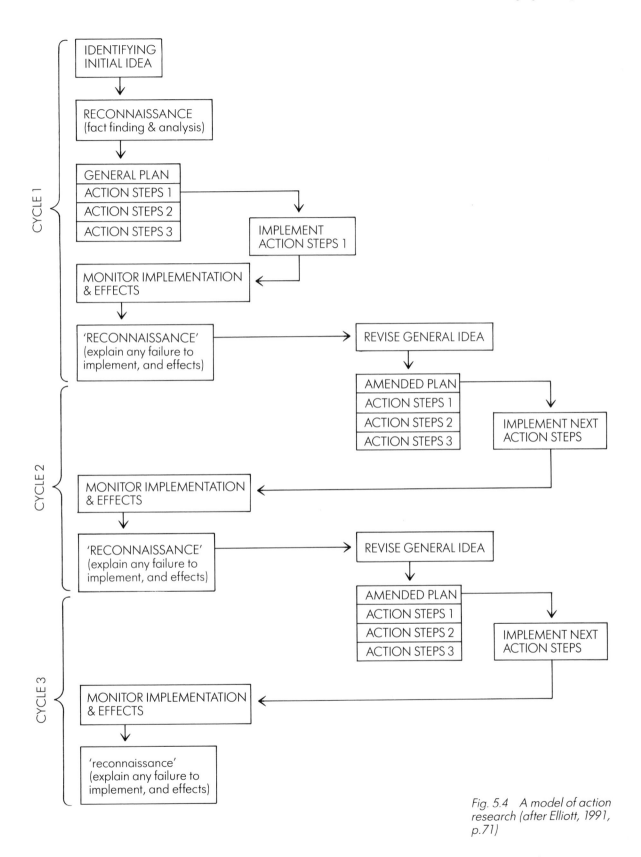

Fig. 5.4   A model of action research (after Elliott, 1991, p.71)

3   Secure rights and gains for the poor

Chambers suggests that

> Sustainable resource use requires that the users take a long-term view. Once their very basic subsistence is assured, poor people's ability to take a long-term view depends on how secure they judge their future rights and gains to be. This aspect of their rationality has been persistently overlooked, perpetuating the myth that the poor are somehow negligently incapable of taking a long view or making long-term investments.
>
> (p.11)

To what extent are similar assumptions made about teachers? It is convenient for both external consultants and internal managers to think in terms of teachers' short-sightedness, their 'resistance to change'. However, teachers may not be irrationally resistant to the changes we would like to see happen; they may simply, like the poor in the Majority World, be quite rationally cautious about taking action which may make their lives more difficult. As Remenyi (1991) says, 'Risk-taking has a uniquely onerous quality for the poor.... The poor are risk-averse because the consequences of failure, should a new technology, product or other innovation not succeed, go beyond the normal financial costs of bankruptcy' (p.23).

Chambers seems to be guiding us towards a crucial shift of attitude. We should assume, he seems to imply, that the clients of our consultancy act with complete rationality – but that that rationality is based on different values, beliefs and goals. It may well, therefore, lead them to extreme caution. Thus our aim should be to offer to our clients some guarantees that the changes we are proposing will secure those things that they value most. When, therefore, we encounter 'resistance to change', we should attempt not to confront it, but to understand its sources, and to offer the security which alone is likely to diminish it.

4   Achieve sustainability by starting with self-help

Sustainability is a key concept in economic and technological development. A project is not sustainable if it collapses without external support and resourcing. Equally, a development in a school is not sustainable if it relies for its impetus on the efforts of a consultant. We all know of cases where developments have relied on a particularly energetic deputy head or head of department, where that person has been promoted on the back of the project's success, and where, within months, all the changes introduced by that person have been reversed by the remaining teachers.

The implication is that sustainable projects are those which rely on the resources of the clients, not of the consultants. As Remenyi says, 'The poor are an asset ... rewards can be improved if the wisdom of experience is sifted and distilled. Much of that wisdom is in the hands of ... the poor themselves, the client group that the [development] programme is intended to assist' (p.120).

The notion of the 'wisdom of experience' is one that is much undervalued in an educational world where teachers seem to have been trusted less and less by central planners. None the less, something of the complexity and subtlety of what teachers know is beginning to be described in a growing body of literature which owes much to the work of Donald Schon (1983, 1990), and it is essential that consultants bear in mind that complexity and subtlety. By so doing, they can avoid the trap of seeing themselves as transmitting to teachers some form of knowledge which is superior to and which will take the place of that which the teachers already have. On the contrary, the focus of any sustainable development has to be not so much the new forms of practice themselves as what Chambers calls 'the enhancement of people's capacity to innovate and adapt, and so to help themselves in the future' (p.12).

In other words, what consultants should be seeking to do is not to substitute teachers' knowledge through traditional INSET processes, but rather to enable teachers to bring to bear on their problems the full weight of what they already know, but perhaps have never articulated. The inevitable corollary of doing this, of course, is that the consultant has to take the risk of freeing teachers from dependency on her or his 'superior' technical knowledge – and it is always difficult to decrease the need other people have for us.

5 The importance of calibre, commitment and continuity of staff

It will be evident from the previous 'lessons' that development consultancy of the sort we are talking about is an essentially human and interactive process. While it demands a certain amount of expertise in the technical sense, it is much more focused on personal qualities. I cannot improve on Chambers' own formulation: 'Calibre refers to sensitivity, insight and competence. Commitment refers to determination, self-sacrifice and dedication to working with and for the [clients]. Continuity refers to working consistently over at least several years' (p.12).

Appropriate technology consultancy in action

What do Chambers' guidelines add up to? At a time when we see educational change as centrally devised and externally imposed, when the freedom of teachers to control their professional lives is diminishing, they suggest a model of change which is essentially internally driven. Change is planned with teachers, to meet the needs of teachers, within the context of the teachers' own culture. While the job of the consultant certainly involves transmitting to teachers whatever external pressures may be relevant, and offering to them whatever technologies seem promising, the control of the process remains with the teachers themselves. As Remenyi puts it: 'More than any other factor, success is linked with the extent to which the client group participates in and even 'owns' the programme and takes responsibility for its health and future progress (p.52).

It would be comfortable at this point to offer readers a set of techniques which they could apply to managing change in their own institution.

However, as Chambers' first lesson tells us, there are no 'blueprints' when it comes to change. The process of consultancy is essentially a process of engaging with particular human beings in a particular situation; the nature of that engagement, the course of its development, and the character of its outcomes simply cannot be laid down in advance. Moreover, whatever techniques a would-be consultant might feel the need to learn, they are of far less significance than the cultural shifts which I have tried to indicate above. Appropriate technology consultancy is ultimately not about the techniques one uses in working with people, but about the stance one takes towards them.

What I propose to do, therefore, is simply to offer a further case study which may perhaps be illustrative of some of Chambers' 'lessons' in action. I leave it to the reader to determine what a similar process might look like in her or his institution:

Fig. 5.5   Case study: developing a new Technology Faculty

I was invited by the Headteacher of a comprehensive school to work with a Humanities Faculty which had been newly formed in response to the projected demands of the National Curriculum. As with many newly formed Technology Faculties, there had been no chance for any real team building, there was no commonly accepted sense of purpose, and many key decisions about the Faculty's practice were being taken at senior management level.

In addition, the Head had very clear views about wanting to 'move the teachers on' from what she saw as a traditionalist, didactic approach towards a more collaborative and differentiated approach to teaching and learning.

When I met the teachers for the first time, it was clear they perceived themselves to be under considerable threat. They knew they were somehow disapproved of, but could not themselves see anything wrong with the teaching methods they were using. Nor could they see what I had to offer them.

Eventually the teachers and I agreed what amounted to a contract. Rather than attempting to teach them new techniques of differentiation or mixed-ability teaching, we would explore together the kinds of teaching that were already going on. The teachers would formulate their own evaluation of that teaching, would identify priorities for development, and would undertake whatever short-term experiments seemed appropriate.

The 'technology' for doing this was remarkably simple. Each teacher described a lesson they had taught recently. Together we identified the principal teaching techniques that they had used in the lesson and displayed the results for the whole Faculty. What they perceived was that, although they had at their disposal a wide range of techniques, and used all of them from time to time, they were actually relying for most of the time on a very narrow range of strategies.

They were, moreover, able to identify for themselves a path of development. They wanted to try to use a wider range of techniques more regularly. They did not need to learn new techniques, simply to make greater use of what they already knew. They would plan some experimentation along these lines, and we would all meet again some months later to see what had happened. In the meantime, they felt that they had uncovered a need to meet together as a whole Faculty, in order to plan together and support one another, and undertook to do this on a regular basis.

It would be misleading to suggest that the time I spent with this Faculty wrought a miraculous transformation. When I returned two terms later for the review session, one teacher complained that they were still, 'stumbling along without facilities to implement an idea that has been perpetrated on us', and they were still not meeting as a full Faculty. However, they did feel that they had changed things: their use of a wider range of teaching styles had increased; they were involving pupils more actively in their learning; and, perhaps most important, six of the eight teachers had decided to involve themselves in an Active Learning Project, in the course of which they met regularly to share ideas and review their teaching. The task for me was no longer to generate enthusiasm, but to harness the enthusiasm of the majority to make sure that they did not leave their colleagues behind. Therefore we spent much of the review session listening carefully to the doubts of those colleagues, and devising ways of evaluating developments on an ongoing basis to investigate those doubts.

The nature of the process

The case study cited above represents one among many forms of involvement with groups of teachers. It cannot be emphasised too strongly that consultants who wish to work in this way must seek the form and format that is peculiarly appropriate to each situation.

None the less, I believe I can detect within the case study elements in the process that are invariably present in some form or other, and that constitute a definition of what I have called 'appropriate technology consultancy'. That definition may help readers think about ways in which they can structure their own work.

Appropriate technology consultancy is the process of working with people in the identification of their needs and wants so that they can learn how to meet those needs and wants in a sustainable manner. This requires the consultant and people to work collaboratively towards:

- understanding and articulating their situations;
- articulating their needs and wants within those situations;
- setting these needs and wants within wider contexts and perspectives;
- exploring means of meeting these needs and wants;
- learning how to implement these means;
- learning how to sustain this whole process independently of the consultant.

What, I believe, makes this process especially relevant to teachers, and above all to teachers of Technology, is that it is essentially a learning process. Certainly, it is concerned with the learning of 'technique'. But it is equally if not more concerned with learning about the situations, needs, wants and values which alone can make technique appropriate. That learning, moreover, is something in which both consultant and 'clients' participate.

The question which inevitably arises for me is whether there is any fundamental difference between the learning process for aid workers and peoples in the Majority World, for teachers and their consultants in the schools of the Minority World, and for those same teachers and their students. Is appropriate technology consultancy, in other words, an appropriate model of the role of the teacher in classrooms and workshops?

I propose simply to raise this question. Although I am fairly clear as to what my own answer might be, I believe it is important that the reader come to his or her own conclusions. However, I think it is worth reflecting on Holly and Southworth's (1989) notion of the 'learning school'. The effectiveness of students' learning, they suggest, cannot be disentangled from the effectiveness of teachers' learning; a school in which students learn is a school in which everyone learns. Too many developments in the education system of late have themselves been profoundly anti-educational. They have been concerned not with teachers' learning, but with teachers' compliance with the will of others. We desperately need an alternative strategy for managing change that will restore the centrality of learning. The notion of appropriate technology has much to offer us.

References

Buxton, James (1983) 'The costly lessons of a $550m dream', in Marilyn Carr (ed.) (1985) *The AT Reader: Theory and Practice in Appropriate Technology*, London: Intermediate Technology Publications.

Chambers, Robert (1988) 'Sustainable rural livelihoods: a key strategy for people, environment and development', in Czech Conroy and Miles Litvinoff (eds) *The Greening of Aid: Sustainable Livelihoods in Practice*, London: Earthscan Publications.

Elliott, John (1991) *Action Research for Educational Change*, Milton Keynes: Open University Press.

Holly, Peter and Southworth, Geoff (1989) *The Developing School*, London: Falmer Press.

Intermediate Technology Education (1992) *Strategies and Guidelines*, Rugby: ITE.

Madeley, John (1991) *When Aid is No Help: How Projects Fail, and How They Could Succeed*, London: Intermediate Technology Publications.

Pacey, Arnold (1983) *The Culture of Technology*, Oxford: Basil Blackwell.

Remenyi, Joe (1991) *Where Credit is Due: Income-generating Programmes for the Poor in Developing Countries*, London: Intermediate Technology Publications.

Schon, Donald A. (1983) *The Reflective Practitioner*, New York: Basic Books.
      (1990) *Educating the Reflective Practitioner*, Oxford: Jossey Bass.

Trainer, Ted (1989) *Developed to Death: Rethinking Third World Development*, London: Green Print.

Chapter 6 The Design Dimension of the Curriculum

Roger Standen

Roger Standen introduces into the discussion the importance of design. Design reflects, most strongly, cultural beliefs and values. Roger argues that we all can participate in the design process, and that design is part of normal human activity, though we rarely categorise what we do as 'design'. He suggests that the whole concept of creating a future that works is, in part, in the hands of the designer, both amateur and professional.

Introduction

In this chapter I outline the 'design dimension', and posit design activity as a basic human activity in which peoples of all cultures and races engage. Design activity, as a capacity which is interdependent with technology, manufacture and human need, is directly concerned with the future and with change. Central to the 'design dimension' is the vital role of general education in maximising potential and development of this inherent life skill.

The key aspect of design activity that I put forward focuses on specific skills which are central to it and provide the foundation for understanding the design motivation on a global scale, whether to provide the simplest survival necessities or to enhance the quality of life within a range of societies and cultures. These are expressed through four basic assumptions about design. They are a touchstone or reference point for teaching and learning about and through design. The range of design skills described include the important cognitive skills of 'imaging' and of 'modelling' which are central to the development of concepts and the communication of ideas.

A number of powerful motivating factors prompt design activity in all cultures and thus affect human survival. It therefore becomes important to reflect upon the effect of design and related technologies on cultural development. The effects of industrialisation are widespread, and the use of technology has affected mass production and commercialism world-wide. It is through an understanding of the 'design dimension' that human beings are enabled to take part actively in the design decisions which affect their own existence and the lives of the wider world community.

Design activity begins early. Very small children display this potential through play, inventing and creating imaginary situations, making and using objects to represent 'real' things, and they interact socially to negotiate and make collective decisions. It makes strong sense to develop this rich potential through education and to create a school curriculum which, from the outset, promotes realistic and inspirational design work in a range of subjects. It is also from this seed bed that the professional designers of the future will come.

We know that design ability is both a complex and specialist process for which professional designers are highly trained. Yet the main point stressed in this chapter is that this ability is also one that everyone possesses. What counts is the degree to which they practise their innate ability and use that fundamental life skill effectively.

Everyone designs: recognising the design dimension of life

Although this chapter is about design as an everyday activity in which everyone engages, the 'design dimension' of living, it does not negate the work of professional designers. It does highlight the valuable contribution of ordinary people to what is so often regarded as an esoteric activity exclusively undertaken by an elitist and talented minority.

A designer conceptualising a cathedral, a new bridge, a ship, garden, irrigation system, townscape, house, book, exhibition or a machine reaches beyond physical problems to imagine new possibilities on a scale which can affect many people's lives. Conversely, and more intimately, families imagine and realise plans for their own lives; for example, to create a home which satisfies physical and spiritual needs. They make their own judgements about the designed places and products they like or dislike, often discussing opinions and challenging others' actions.

In all societies people develop a deep awareness of their own cultural values and qualities: they are moved and excited in varying degrees by things that others have made. They seek them out in shops, markets, museums, galleries, temples and religious buildings, trying to discover their human meaning.

All human beings are both full of curiosity and creativity, and so speculation about the future has been, and still is, the key to a powerful motivation for design activity. To reach out and extend the scope of human achievement is one of the most powerful motives for design activity and technological development.

Historical study reveals warfare and economic competition as two main driving forces in technological innovation and development. In a competitive and changing world they will continue to make demands, needing vast resources of human skill and raw materials.

Industrialisation and new technologies have increased the complexity of the made world and, in consequence, interaction between environments, products and communications. As the effects and side effects of technological innovation become more far-reaching, so the task of using technology wisely to serve human needs and desires becomes more demanding.

Against this background, education about and through design and technology needs to develop as a central concern. We should also consider carefully the potential effect of Western technologies and cultural influence on all societies and cultures. Important qualitative differences exist between those dependent mainly on craft-based means

An advertisement for a Fiat car 'Designed for Life', 1991!

of production and those where automation and mass production are accepted as a normal part of life and values.

Gandhi said that what India needed was production by the masses, not mass production. This reveals a difference in cultural, philosophical and commercial development between mass-producing cultures and largely craft-based ones. It is not right to assume, however, that because sophisticated, mass-produced technology may provide for many people's needs it is superior to one that does not. The importance of appropriate technologies, production, commercialism, the creation of wealth and the spiritual well-being of communities are also aspects to consider.

The central issue of needs and wants allows us to reflect upon the moral implications in designing. Later in the chapter, in outlining the basic assumptions about the 'design dimension', this aspect will be explored more fully.

'Wants' leads to 'needs'

The sophisticated techniques used by advertisers not only give information, they also deliberately manipulate text, sound and images (see photo above). These are powerful tools, which can be used to create needs out of wants. Consider the selling of any item which is not required for basic survival; the motive must surely be to generate the desire in potential consumers to obtain what is on offer. Part of people's good design understanding is an ability to discriminate, to decide to accept or to reject, to distinguish between that need or the desire to own.

Culture and values, also featured in the basic assumptions, challenge the neutrality of design, particularly in emotive areas such as religion. What has been the relationship between the creation of temples, mosques, churches or cathedrals – some of the most significant of the world's architectural and engineering constructions – and the economic, social, technical, moral and political skills used in their design and

The Vatican, Rome – what has been the relationship between the creation of some of the world's most significant architectural constructions and the economic, social, technical, moral and political skills used in their design and construction?

construction process? Consider too, how the special products and artefacts that religions communities use address in their design the issues of needs and wants. Quite clearly, philosophy and values need to play a very significant role in understanding and resolving these questions.

Basic assumptions about design

Designing: a fundamental capacity of all human beings

The first basic assumption is that designing is a fundamental capacity of all human beings, and that it is not separate from the general run of human concerns but an intimate and normal part of them. It is inherent in the way we learn, the majority of actions we take, in the ways we organise and manage our lives, in the values and beliefs we hold, and in the places we create and the products and services we make and use. Designing is the work of deciding how places and products will be made, what functions they will perform and what they will look like in the future. It is the outcomes from this activity that provide a way of recognising cultural diversity, of the individuals, groups or societies creating the products and environments.

Design awareness and understanding enables all people to enjoy and comprehend with insight their made world, and to take part in personal and public design decisions which affect their lives. It encourages critical consciousness of quality and of the qualities of designed products and environments.

Design activity is concerned with the future

The second basic assumption about design activity is that it is primarily concerned with the future and how people will live. Central to this is an understanding of how environments and products of the present and the past have been created, as, embodied in them are cultural traditions, knowledge, experiences and expertise reflected in the customs and civilisation of those peoples.

Values and culture

When considering carefully the question, how do we want to live in the future?, what is decided and the ensuing action may largely be determined by value judgements.

The third basic assumption concerns values and culture. Both are interlinked and embedded in design thinking processes and methodology. Designed things often reflect the values of their designers.

What is needed for the future? What do people want? By what means could it be achieved and at what cost? How will what has been planned affect people, their environment now and in the future? Values and value judgements always emerge when striving for the best balance or proposing the most inclusive and potentially enhancing range of possibilities. People think about and make design decisions, they take considered actions and create new objects whilst reconciling conflicts and compromising differences in the process. Evidence in the made environment reveals the influence of many pressures.

In the second section of the chapter I referred to people's innate sense of curiosity, a need to make sense of their world and to appreciate and be inspired by things others have made and the buildings and places they have created. Tourism enables many people to have these experiences, and the growth of this industry both gives rise to many opportunities, and also impacts upon the people and places targeted. Consider the effect of tourism on a craft-based society where the desire for souvenirs has kept the traditional craft skills alive and supported the economy of those people. In what ways might tourism affect the future for those people?

This is one of the most important questions, for their development depends upon it. We should ask whether they are developing products which will have true value in the future in the same way as those of the past are valued now. How will the traditional codes of behaviour in dress and as regards society as a whole be preserved?

The moral dimensions of design are evident in all cultures, too, with potential repercussions at personal and regional level. Design is not morally neutral, it is, in fact, morally problematic; it throws into practical relief a range of good and bad human motivations. It is as much concerned with shaping a war memorial as with creating a pacifist film. The made environment reflects back all too clearly an image of people's values – greed as well as altruism, corruption as well as honesty; competition as well as cooperation and, frequently, profit-making above all else.

Earlier in the chapter needs and wants were mentioned with reference to the moral dimensions of designing. The role of advertising and the designer's part in it is an interesting one. Most of us are well able to recognise advertisers' stereotypical portrayals of ideal families, fast, glamorous life styles of endless fun, good health and prosperity, and to accept them for what they are. Others, less aware, need to become more conscious of ulterior motives. It is when the advertised product itself is morally problematic that serious moral dilemmas come to the fore and questions arise. Do the designers creating an advertisement to sell tobacco reflect honestly on the moral implications of their actions? Can it be right to promote a product which is known to kill huge numbers of human beings? Is it even worse to promote such a product in countries where tobacco abuse has been, hitherto, largely unknown and by so doing initiate a series of directly related problems?

Visual imagery plays a large part in the giving of information, and people readily respond to the messages of advertising, film, television, signs and buildings. Moral questions too are associated with visual imagery. An example of this is the power of religious visual imagery. Consider how successful the symbolic devices are and how they have been used to express meaning, to inform, to tell a story or to explain moral codes, or even blatantly to exercise propaganda.

Design skills

The fourth basic assumption concerns the extensive range of design skills which are both directly concerned with the act of designing and the iterative process. The skills described here are not exhaustive; they interact, for designing is neither a simplistic nor a straightforward activity. Economic, social, technical, aesthetic, moral and political skills are linked to those perceptual, analytical, propositional, communicatory, technical and manual skills. They are skills based on acquiring information, understanding and knowledge of situations and the means by which human beings present, negotiate and illustrate ideas.

The vitally important aspect of design activity is the acquisition, development and communication of design concepts and ideas.

Imaging

The unique human ability to image is a cognitive skill which enables people to imagine the world differently. This imaging skill enables us to envisage and to plan how, for example, our house or garden, field or yard might look in the future. However, imaging can only be a really useful skill if what people 'image' can be successfully communicated to others.

A way of doing this is to devise a series of symbols to represent the ideas. In the design sense these symbols are known as 'models', for they represent reality. Amongst the many kinds of models are diagrams, notes, sketches, plans, drawings, scale plans, mock-ups and prototypes – all of which can help to communicate 'imaged' possibilities and foster skills in using such media to assist thought and promote action.

'The home I would like to live in' – a painting of an imaginary house that can walk on water, by a five year old

Planning the future – deciding today what will grow next year. Garden design is a good example of our ability to imagine what something might be like and making the best uses of available resources

Selection of the most appropriate device is critical and maximises the potential for negotiation and clarification of ideas. A scale plan of a site for housing provides an example of how, with representations of the proposed houses and perhaps the existing trees and other natural features, the best arrangement of parts can be tried out and negotiated before the design is finally put into practice.

People use these imaging skills constantly and in tandem with other skills; for example, economic ones. Not solely concerned with finance, these are about making the best use of available resources, adding or giving value to raw materials, to ideas, ideals, philosophies or beliefs.

Social skills are evident in the ways societies are structured and social customs and traditions operate. The influence is evident world-wide in house designs, layout of public spaces, clothing and public buildings. These skills are much concerned also with how human beings interact, live and work together in social groups. People's communication and social skills are interlinked, especially in presenting proposals and ideas and in negotiation. Aesthetic skills are integral to quality, design, work and process.

Principles and understanding must be applied in designing objects or environments or in communication because, without aesthetic understanding, people are unable to respond effectively and sensitively to materials, structures, form, shape and function and to make sense of space, proportion, scale and the use of colour and texture in designed things. It is through a rigorous understanding of aesthetic qualities that greater appreciation of the needs and wants of other nations and cultures is achieved.

Political skills in a design context refer to those skills needed for developing strategies or procedures to achieve a particular goal. People's actions are sometimes governed by 'political' awareness. Business practice often employs such strategies which require delicate manoeuvre and negotiation, for example, so as to secure contracts.

Last to be mentioned are technical and manual skills. These are the skills which enable people to create artefacts and environments and communications. Peoples of all cultures use tools and materials purposefully to build, to make clothes and utensils and to process their food.

Conclusion

Design activity has been put forward, throughout the chapter, as a fundamental capacity of human beings. This design activity, which decides the form that things will take, is interdependent with technology, manufacture and human needs.

The form of things cannot be considered apart from the purpose given to them by human beings. Human purpose, however, does not, as we have seen, have free rein; achieving it depends on the resources available to construct what is needed and the available knowledge about making things work in the physical world. Design and designing in the broadest sense consists of a bundle of technologies, skills and approaches which are instrumental in determining the future.

The City of London – the past, present and future. How do I want to live in the future?

When we refer to the 'design dimension' in education and recognise it as an essential part of the school curriculum, we should put 'design' and 'designing' together – 'design' to indicate the qualities that result from having been designed, and 'designing' to indicate the activity of engaging in design. The distinction has been between the physical form that designs take in the outside world and the human ideas, values and actions that are used to determine or create these qualities. Any educational design activity needs to embrace both these perspectives.

Industrialisation and technology are ubiquitous; they have vastly altered the quality of ordinary people's lives. In democratic societies people have the opportunity to have an effective say in the important decisions that will determine the future pattern of social life. To do this education has a crucial role in bringing these issues to the fore, and in challenging them and developing ways of study and making valid contributions by the non-specialist, through learning and developing design awareness.

The ability to handle change in a purposeful way has been highlighted for individual and social survival, in cultures affected by tourism or other impacts. Mass production has to some extent divided the consumer from the market, but the consumer still needs to exercise a broad range of skills and understanding so as to control fully or change his or her environment. An experience of design and designing will help people return a personally valid answer to the question, 'how do I want to live in the future?'

This is a universal question and is the driving force that underpins the development of all societies, whether at present part of the Minority or Majority World cultures.

Part II Case Studies, Activities and Illustrations

Chapter 7 Global Perspectives in Technology Education

Sue Greig

Sue Greig has picked up many of the points made in the first chapter. Development Education has not found much of a home in Technology: now there are real opportunities for a global perspective. Her chapter introduces Part II of the book, which has examples of classroom activities. Her chapter sets the scene for that more practical focus, in a clear way, drawing on the framework established in Part I. She describes some possible projects, pointing out their appropriateness for National Curriculum Technology, and encourages the teacher to tackle some of the crucial questions facing our world.

The Technological imperative

Technology – that is, the making of artefacts and systems to solve problems – is inevitably a product of the values and cultures which identify and prioritise those problems and which evaluate possible solutions.

Over the last two to three hundred years science and technology has developed largely in a culture which has viewed nature and the natural world as separate and inanimate, rather than as a dynamic system in which we play an integral part. With nature seen as a collection of objects to be dissected, manipulated and controlled, science and technology have achieved a spurious detachment from other areas of human thought and activity. Thus the vested interests and ideologies which have shaped scientific and technological development for their own ends have been obscured beneath a veneer of neutrality. In this context, and with a momentum all of its own, technology has pursued narrow goals, without regard to wider human and environmental impacts. In the words of Lewis Mumford,

> Western society has accepted as unquestionable a technological imperative that is quite as arbitrary as the most primitive taboo: not merely the duty to foster invention and constantly to create technological novelties, but equally the duty to surrender to these novelties unconditionally, just because they are offered, without respect to their human consequences.

The 'technological imperative' has several elements. One is a fascination with our own cleverness; another more important element is that there is money, power and influence to be gained by the continual creation of 'technological novelties'.

Technology has certainly achieved benefits, but they are simply not available to the majority of the world's people. It has also incurred heavy costs. Much technological effort is directed at destroying rather than enhancing life, at increasing consumption (and pollution) rather than resource conservation. Technology currently serves neither the majority of the world's people nor the planet on which we all depend for survival. Some might say the problem here is not technology, but politics or economics or ecology. However, such compartmentalisation is a further reflection of the problem, rather than part of the solution. We need to put the politics, economics and ecology – that is to say, the people and the planet – back into technology, and crucially, back into Technology education.

Global education and technology

Initiatives in global education, and in the related areas of development, environmental and multi-cultural education are all attempts to make the school system more relevant and more responsive to the fast-changing and interdependent world we now live in, and to the uncertain future that those now in our schools will face. Several writers have attempted to tease out the definitive elements of a global perspective in the learning process*. The four strands described below can all be seen to be central to a broader, more global view of technology education.

Systems consciousness

A broad view of technological activity considers the maker in relationship to other people, as well as to that which is being made. It considers the raw materials and the waste products, and future and distant impacts, as well as those in the here and now. It recognises the need to consider the system as a whole – so often, narrow-focus technological solutions solve one problem but create several more in the process.

Perspective consciousness

We all have a world view which is not universally shared. Values and experiences differ, and therefore different people will identify different problems and opportunities for technological activity and will develop different solutions. A broad view recognises and values such differences and attempts to empathise with the situations of others.

Health of planet awareness

An understanding of the current 'state of the world' and of development and environmental issues and trends is vital to an understanding of technological impacts in the world.

Human choice and responsibility

We are all involved in our everyday activities in choices and decisions about the use of technology. Conflicts of interest can and do arise. The controversial nature of technological activity is central to a broad view of Technology education.

* Pike, G. and Selby, D. (1988): *Global Teacher, Global Learner*, London: Hodder & Stoughton, pp. 34–5.

Table 7.1   Comparing narrow and broad focus Technology education

Narrow focus	Broad focus
1 technology for its own sake	1 technology as a tool for sustainable development
2 emphasis on product as object	2 product seen as part of a system, of producer(s), product user(s), raw materials and waste products
3 technology for experts	3 technology as an everyday activity
4 sophistication as goal	4 appropriateness as goal
5 ethnocentricity	5 draws on a wide range of cultures and values
6 claims technology to be neutral or value-free	6 acknowledges controversiality of technological activity.

The medium is the message

Broadening the content of Technology education is clearly part of the goal, but the learning process is also crucial. For students to develop the dimensions of a global perspective outlined above, a high level of interaction in the classroom is essential. Tasks should be structured so that students work in pairs or small groups, to encourage discussion, to air alternative perspectives, to negotiate and to make decisions. In short, the aim is to encourage cooperation and participation by all. Students are offered opportunities to examine their own points of view and to attempt to understand those of others through the use of role-play and simulations. Attitudes and values have as much place as 'factual' knowledge and skills.

Ideas into action

The following activities aim to show how the ideas outlined above may be translated into classroom practice. The context offered is that of National Curriculum Technology, but the ideas and techniques are readily transferable to other areas of the curriculum.

1   Understanding contexts; identifying needs and opportunities

Purpose
In the following activity groups are each given the same problem to address, but within a range of differing contexts.

Procedure
Small groups of three or four are each given a picture of a different type of building. Their task is to consider how they would go about designing a heating system for the building in question (Figure 7.1).

They stick the picture in the middle of a large sheet of paper and write down around it all the questions they think they would like to ask about the building (Figure 7.2).

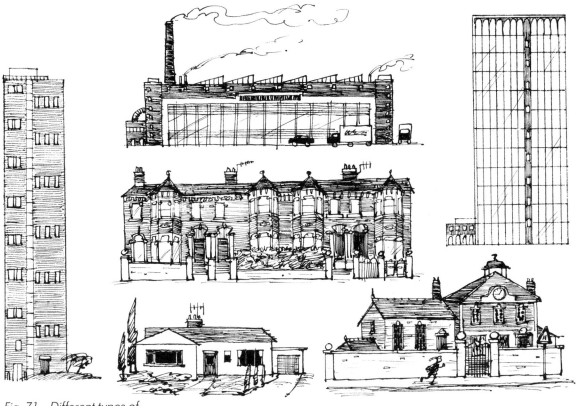

Fig. 7.1 Different types of
buildings

What do the surroundings
look like?

How many people use the building?

What type of fuel is available?

What is the building
used for?

How will the fuel
be transported?

Where will the fuel
come from?

How much money is there
to put the heating system in?

Who will pay to run
the heating system?

Fig. 7.2 What do we need to
ask when looking at
buildings?

The sheets are then displayed around the classroom and groups are given time to circulate and look at one another's work. Members of each group take it in turns to stay with their sheet, to discuss any points with others.

Potential

As students frame the questions they need to ask, and then share them with other groups, they gain a deeper understanding of the influence of the context on needs and opportunities for technological activity. In particular, questions that were asked in some contexts, but not in others, can reveal assumptions which, up to that point, students were unaware that they had made. For example, did they assume that they knew what would be going on in the building? What about the building's surroundings? Did they assume that the building was in the United Kingdom? If it was elsewhere in the world, are there other questions they would need to ask?

2   Developing and communicating ideas: generating a design*

Purpose

The aim of the following drama-based activity is for students to work together in small groups to develop a cooperative design. The theme chosen is that of recycling, but the basic idea could be applied to a wide range of contexts.

Procedure

Some work on recycling, in particular on the nature of the cycles involved in recycling different materials, would be useful preparation for this activity, although not essential.

In groups of four or five, students are asked to create an imaginary recycling machine that they can demonstrate to the rest of the class. They can choose whatever material they like (even resources that cannot, at the moment, be recycled).

First, students decide which waste material they would like to consider for recycling. When they have agreed on a material, they think about how to create an imaginary machine for recycling their waste into something usable.

Each group should discuss the following questions:

1 What will the waste material be used for?
2 How big will the machine be?
3 How many different stages will there be in the cycle?
4 How will the machine be operated and what will it run on?
5 How many people will be needed to operate it?

Students work on finding a way of demonstrating their machine to the rest of the class. This can be with students either becoming parts of the machine,** or becoming the people that operate it, or both.

* This exercise is based on classroom work developed by Karen Coulthard, Headteacher, Berger Junior School, London, and is described in *Green Teacher*, issue 10, 1988.

** This kind of activity is expanded in Chapter 10.

Making something useful from a leaf spring, Zimbabwe

Then students imagine that they are working the machine. What sort of things will they be doing? What might the people operating the machine talk about?

Each group shows their machine working to the rest of the class for two or three minutes. They also describe a little of their working day. The rest of the class are encouraged to watch carefully and then to ask questions and comment on what they have seen and heard. Questions may be about the machine itself, the resource that is being recycled, or how the group reached their final design.

As a follow-up, each group could draw a diagram of their machine, incorporating how they think the whole recycling process may work (see Figure 7.3).

Potential
The activity will help development of cooperative and hypothesising skills. Although the machines that are designed might not be at all realistic, speculation like this on an imaginary level allows students to question their own knowledge, which can be extended through sharing their ideas with the rest of the class. The activity gives them an opportunity to explore what recycling a material involves, and can lead to further research into what is practical. It also offers the teacher insights into her students' own understanding of the issues, on which subsequent work can build.

3   Using appropriate resources: planning and making

Purpose
This practical activity allows students to investigate for themselves a soil and water conservation technique which is being used very successfully in the African Sahel, with the aid of a simple surveying tool.*

Procedure
The principle behind water-directing techniques and the Sahelian context is explained to students, using the following background information.

Context
In the early 1980s Oxfam workers took a number of simple soil and water conservation techniques, which had been developed in desert regions of Israel, to the north of Burkina Faso in the African Sahel. All the techniques revolve around focusing scarce water on places where plantings have been made. For example, small basins dug around tree seedlings will improve their chances of survival. But African farmers were more interested in improving the survival and yields of their food crops, mostly millet. Traditionally in this area, the farmers clear their fields of the many heavy ironstone rocks found on the surface, and they line these rocks along the edges of the fields. This was adapted into a technique called 'contour damming'. Very long, very low ridges (15–25 cm high) of soil and stones (or sticks) are built at right angles to the slope, and these stop rain-water run-off, putting water into the soil. If the traditional stone lines were built more solidly and aligned with the contours, using a simple surveying tool, they were ideal for the purpose. The technique has been taken up enthusiastically by the farmers, and has spread to more than 300 villages. The stone lines take a lot of time and effort to build, but they have resulted in much-improved grain harvests, and have been used to bring long-abandoned land back into cultivation.

Part 1: contour damming   This part of the activity can either be done outside or simulated in the classroom, using large trays filled with soil. If the activity is to be outside you will need a gently sloping area of bare soil, divided into enough areas for groups of up to six to work on. In the classroom, you will need one large tray filled with soil for each group of up to six students. The soil-filled trays should be propped up inside a

* This activity is approached in a different way in Chapter 10.

The Pipe industries

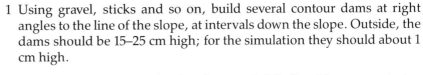

Fig. 7.3 A child's drawing of a recycling system

larger tray to simulate a gentle slope (see Figure 7.4). Students are given the following instructions:

1 Using gravel, sticks and so on, build several contour dams at right angles to the line of the slope, at intervals down the slope. Outside, the dams should be 15–25 cm high; for the simulation they should about 1 cm high.

2 For ten minutes or so, simulate heavy rainfall. Outside use a watering can, and inside use a plastic sweet jar with about thirty holes in the base.

3 Watch carefully what happens, and then try to answer the following questions:
What effect do the dams have on the way the water flows down the slope?
Are there any places where the water has broken through a dam? Why do you think this has happened?

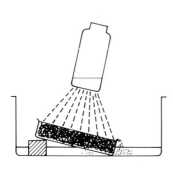

Fig. 7.4 Testing contour damming

Part 2: improving the system   There will be some places where the dams do not go exactly along the contours. This will mean that water concentrates on one part of the dam and breaks through, running off down the slope. It is to be hoped that students will notice this and work out the reason for themselves! If not, it should be pointed out to them.

Then they are given the task of finding a way to build dams exactly along the contours, even when slopes are very gentle indeed and contours are impossible to judge by eye.

Each group is given two wooden stakes and a length of clear plastic pipe, and asked to construct a tool which will enable them to improve their contour damming system.

Once they have worked out how to construct the simple hosepipe water level shown in Figure 7.5, and if they are able to work outside, they can experiment with it and try to build an improved contour damming system.

Students are asked to think about why this technique has been put into practice so enthusiastically by the farmers in Burkina Faso and in Turkana, in Kenya, even though building these dams takes a lot of time and effort. Various other solutions might have been found – from large-scale irrigation projects to bringing in surveyors and sophisticated surveying equipment to tell the farmers where their stone lines should be built. Students are asked to think about and discuss whether other approaches would have been as successful, and if not, why not?

Potential

Through this activity, students gain direct practical experience of appropriate technology in action.

Fig. 7.5   Using a hosepipe to improve the damming system

4   Meeting needs: evaluating technological solutions*

Purpose

The following role-play activity aims to enable students to understand the conflicts of interest which can arise as a result of technological innovation, and how such conflicts can lead to outcomes quite different from those envisaged.

Procedure

The bio-gas project is explained to the class, using the background information provided. The class is then divided into small groups of three or four. Half the groups are labelled 'A' and the other half 'B'. Each group is given a copy of the promotional material on methane generators, and information about their own situation (A or B). It is explained that after 10 or 15 minutes they will have the opportunity to speak to a representative from the voluntary agency which is promoting the bio-gas project, and to say what they think about it.

Each A and B group reports to the development agency representative (played by the teacher) on their view of the proposed project.

The class as a whole then consider the costs and benefits of the bio-gas project for the two groups of villagers (namely, those with and without cattle), and for the environment. Will the bio-gas project help to save India's forests as it is designed to do? The discussion could be summarised in a table on the blackboard or flip chart.

In small groups of three or four, students then discuss ways in which bio-gas technology could be used for the benefit of all the villagers and for the environment.

The small groups share their ideas with the rest of the class.

Potential

Through this activity students may understand better the need for looking at the wider picture when assessing the possible impact of technological innovations.

Bio-gas for the village: background information

In many parts of India, increasing population and a greater use of land for agriculture has put pressure on the forests. In particular, the villagers' need for firewood for cooking has been causing deep concern amongst conservationists for many years. One development that has been supported by many voluntary agencies and by the Indian government is the use of methane generators which can turn animal wastes into clean and cheap gas which is ideal for cooking.

Many different versions of the bio-gas plant have been produced, some small and some large, some using steel and brick whilst others have been made out of fibreglass. The whole technology has attracted an enormous amount of engineering effort as well as mobilisation of villagers to adopt the new technology. Although different experts continue to favour one design or another, it is generally agreed that bio-gas technology has many economic and ecological advantages. Never the less only those with a

---

* This exercise is based on a real life account: see Taylor, L. and Jenkins, P. (1988) *Time to Listen*, London: Intermediate Technology Publications, p.4.

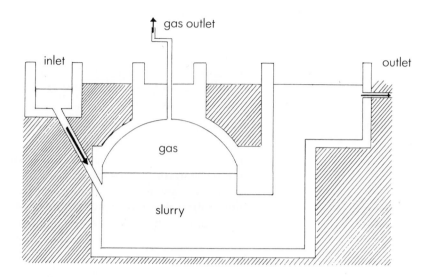

inlet

gas outlet

outlet

gas

slurry

Fig. 7.6   *A methane generator*

certain amount of savings and a certain minimum number of cattle can really benefit from this new development.

(L. Taylor and P. Jenkins (1988) *Time to Listen*, Intermediate Technology Publications)

Group A

You have your own cattle and a small amount of savings – enough to buy a methane generator. At the moment your cattle are free to roam around the village; you have no use for the cow dung they produce. You use firewood for cooking which you collect from the local forest, although you are having to go further and further to collect enough. If you have any extra firewood, you make it into charcoal which you can sell at a good price in the local town.

1 Do you think the bio-gas project could improve life in the village for you?
2 What are you going to say to the development agency representative?

METHANE GENERATORS
Special Introductory Offer
Turn your cow, sheep or goat dung into clean, cheap fuel

Ideal for cooking

Simple to use

Save our forests – use Bio-gas

Fig. 7.7   *Promoting bio-gas*

Group B

You do not have any cattle of your own, and you do not have any savings so you cannot afford to buy a methane generator. Other people in the village do have cattle, and they let them roam freely around the village. You collect any cow dung which you find, and dry it so that you can use if for fuel to cook with. Sometimes you use it to fertilise your small plot of land where you grow food; if you collect a lot of cow dung you can sometimes sell it to other people.

1 Do you think the bio-gas project will improve life in the village for you?
2 What are you going to say to the development agency representative?

5   Information technology: what is development

Purpose

Students use information technology to look at the correlation between a typical index of development: Gross National Product and energy consumption. They go on to consider where development, defined by economic growth, has taken us and is likely to take us in the future.

Procedure

A good introduction to this exercise would be to brainstorm with the class their ideas on the meaning of 'development'. The following definition of GNP is given to students to consider. The total wealth a country produces in the course of a year is called its Gross National Product. Wealth produced means goods (baked beans, shoes, soap, toys, computers, and so on) as well as services (transport, banking, health services, education, waste disposal, crime prevention, among other things). To calculate GNP, the value of all goods and services produced in a year are added together. GNP per capita is GNP divided by the total population. It shows what each citizen of a country would receive if the country's wealth was divided equally.

Students then use the data below (and, ideally, a computer) to complete the following tasks, in pairs or small groups.

1 Produce a plot of energy consumption per capita (on the y axis) against GNP per capita.
2 What difference do you notice between the Minority World countries and those in the Majority World? Read the definition of GNP again; can you explain the differences? Do any of the data for individual countries surprise you?
3 If all the countries of the Majority World reached a similar level of GNP per capita to those in the Minority World, what would you expect to happen to energy consumption around the world? Do you think countries will be able to continue increasing their GNPs year after year? Explain your answer.
4 Do you think GNP is a good measure of how developed a country is? What other measures could be used? In your group, discuss and then agree on words to complete the following: 'A developed country is one where ....'. According to your agreed description, is Britain a developed country?

Table 7.2   *Wealth and energy consumption around the world, 1986*

Country	GNP per capita ($US)	Energy consumption per capita (gigajoules)
Algeria	2,570	37
Burkina Faso	150	1
Egypt	760	21
Ghana	390	3
Nigeria	640	5
South Africa	1,800	81
Zimbabwe	620	19
Canada	14,100	284
El Salvador	820	5
Jamaica	880	32
Mexico	1,850	47
United States	17,500	278
Argentina	2,350	51
Brazil	1,810	22
Peru	1,130	15
Bangladesh	160	2
China	300	21
India	270	8
Israel	6,210	75
Japan	12,850	106
Malaysia	1,850	31
Saudi Arabia	6,930	122
France	10,740	114
Hungary	2,010	109
Italy	8,670	94
Poland	2,070	138
United Kingdom	8,920	157

Potential

There are several commercial software packages available concerned with development data. The World Development Database (available from Centre for World Development Education) contains files on those 125 countries which have a population of more than one million with statistics on thirteen social and economic indicators, including life expectancy, adult literacy rate, population growth and GNP per capita. The files have been prepared on GRASS and QUEST information handling systems.

Educating for change

People tend to have one of three attitudes to technology:

1 technology holds the key to all our problems;
2 technology is the cause of all our problems;
3 perhaps the most widespread attitude is indifference: 'technology? – nothing to do with me, I leave that to the experts!'

None of these attitudes is likely to bring about change for the better, and as educators we should not let such attitudes go unchallenged. We must expose the shortcomings of technological 'progress' which leaves millions of fellow human beings going to bed hungry each night, with no access to clean water and shelter, whilst the support systems of the planet become ever more degraded, through the activities of a minority. We must broaden the scope of technology education, to include the wider human and environmental impacts, and to explore ways in which technology can be a tool for sustainable development. (Here we will often need to look to the Majority World for positive examples.) Perhaps most important of all, we must engage young people in a debate on technology, development and the future, which acknowledges that all of us, in our daily lives as citizens, as consumers and as workers, make choices about the use and development of technology. We cannot be uninvolved. That is what Technology education with a global perspective is all about.

Chapter 8 Appropriate Designing Begins at Home

Ann MacGarry

This chapter should put paid to the idea that appropriate technology is 'third rate' for third world. Ann MacGarry has given teachers a wealth of practical starting points, and has put her own perspective on the underlying issues, which come up elsewhere in the book. The Centre for Alternative Technology, in Wales – where Ann works – provides teachers and their pupils with many opportunities to examine and experience what alternative, or appropriate technology really means, and what relevance alternative technology has for our own lives. Her unthreatening style should persuade teachers that they can tackle these issues, and that they can provide contexts that have real meaning, particularly with the aim of making the future work.

Many people from the Majority World are very suspicious of the idea of intermediate or appropriate technologies. They feel it is just another way in which they are fobbed off with the second- or third-rate things which we, in industrialised countries, would not dream of using ourselves. If they visit Britain they do not usually see any evidence of our using the same approach to technology that they feel is being preached to them. Perhaps we should be using appropriate technologies more widely.

What makes a technology appropriate?

What happens, for a start, if we look at a list of criteria for what makes a technology appropriate in the Majority World and at how each factor applies in Britain? When this is examined in school the pupils should, of course, draw up the list they have arrived at themselves through discussion, but, as an example, here is one used by Intermediate Technology Education to promote discussion – there are no 'right' or 'wrong' answers.

1 A technology that best suits the needs and lifestyles of the people using it
What is a need? We have to explore with pupils the difference between needs and wants or desires. Can any society afford any longer to focus on satisfying wants rather than needs?

Can the world afford the consumption of materials and energy that this involves? Do we know what we need any more?

Most of us think that we need our own private car, but this causes pollution, damages the environment and therefore threatens our future, and also undermines our health by making us less fit. Some people must even think they need one of those electrical gadgets that takes the fluff off jumpers.

A useful class activity would be to look at some of the advertising catalogues which often come with the Sunday newspapers, and identify what needs (!) or wants (!) are being met by the items for sale.

2 Non-violent to the environment and ecosystem

We in the Minority World are doing far more damage to the environment and ecosystem than are people in the Majority World, and it is therefore even more important that we reduce this damage. As well as trying to save the Brazilian rainforest perhaps we should be replanting the forests that we cut down in the course of our industrial and agricultural 'progress'.

3 Costs should be within the economic means of a community

When it comes to looking at costs, one needs to differentiate between the here-and-now financial costs and what I would call the 'real costs'. Both of these must be considered, but we in the Minority World rarely pay the real cost of what we consume.

The real costs would include:

- the financial cost of continuing to operate the technology in the future;
- the environmental cost (that is, for example, the contribution to global warming) of using energy from fossil or nuclear fuels in extraction, processing, transporting and packaging;
- the cost to future generations of using up scarce and useful raw materials and energy resources;
- the cost of cleaning up the pollution and disposing of waste caused in processing (soil, water and air pollution and industrial or agricultural waste products);
- the human cost to all those involved in production.

What are human costs?

Many things that we use (coffee, aluminium, uranium, copper, electronic components, chocolate) are mined, grown or manufactured by people who work for very low wages, often in conditions that damage their health, because their only alternative is even greater poverty. Some of these products are grown on fertile agricultural land in countries where many people cannot get land or jobs and therefore have inadequate diets. Many of these products are nutritionally valueless and even harmful, like tobacco or coffee. (Do not, however, assume that all people who work long hours in bad conditions for low wages are 'over there' on the other side of the world.)

This is all part of appropriate economics* and is in the context of our use of these materials. But how does one cost the fact that good agricultural land is used to produce a 'food' like cocoa which is very pleasant but nutritionally of minimal value? The prices of raw materials like cocoa on the world market have declined steadily over the last twenty years in real terms. ('In real terms' means that the producers can buy less for their money.) The prices of oil and manufactured goods from the Minority World do increase in real terms. For example, in December 1990 the price of bananas on the world market was only 15 per cent of what it was in 1965. This means that in 1990 a producing country had to sell six to seven times the number of bananas it sold in 1965 in order to buy the same number of vital things like trucks.

If you look at all commodity prices (not including oil) in 1989 they were only two-thirds of what they were in 1980.

* This links with Ken Webster's thesis in Chapter 4.

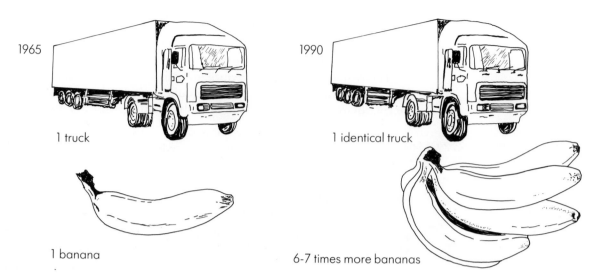

1965 — 1 truck
1 banana

1990 — 1 identical truck
6-7 times more bananas

Fig. 8.1   In December 1991 the price of bananas on the world market was 15 per cent of what it had been in 1965

The real cost of treats like chocolate would include paying a wage to the growers, pickers and processors that was equivalent to an average wage in Britain, and a price for processing and transporting that took into account the use of diminishing world resources and pollution. If we had to pay this cost then we would consume a lot less and the item would again be seen as the luxury it really is.**

4 (a) Locally available materials should be used as far as possible
Why should we do this in Britain? Or what are the financial and environmental costs of not doing so? Let us look at something which uses imported materials, taking as an example the most famous product of the fast food industry, the beefburger. The beef has probably been imported frozen, the roll is probably made from Canadian wheat and the tomato sauce has very likely been made from Italian tomatoes. The onions might be British but they might well be Spanish. What is the cost of transporting these things across the world? What has one miserable little beefburger contributed to global warming? The economic and environmental costs of transporting all these things, but particularly the frozen beef, are very high. Most or all of the energy consumption will involve burning fossil fuels.

Fig. 8.2   Where does a beefburger come from?

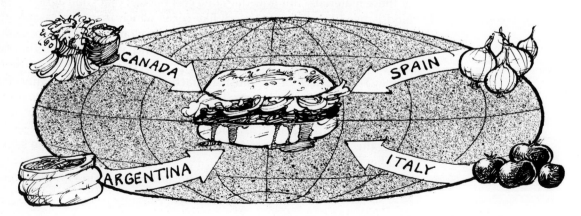

CANADA   SPAIN   ARGENTINA   ITALY

** Bernard Guri's comments about chocolate are relevant here (see page 23).

It is very interesting to look at all the things we buy in a week, not just food, and consider how far they have all travelled. Was this necessary and what did it cost in terms of the energy used for fuel and refrigeration? Do we really need out-of-season green beans, flown in from Kenya? It would be a useful exercise for pupils to make a list of all the things bought by their household in a week and draw up a chart like the one below.

Object	Need/Want	Material	Country of origin	Refrigerated	Tinned	Packaging	Energy use

Then they could consider which of these things needed to be imported. Sometimes it is more energy-efficient (and therefore probably more appropriate) to import a particular material than to use an alternative, locally available one. There are no hard and fast rules about these things. Every situation is different.

4 (b) Manufacture of the technology should be local
Not only would this mean that energy would be saved in transportation, but it also encourages local employment and ensures that repair is likely to be easier.

5 The technology should enable local artisans to earn a living
In a country where unemployment is an increasing problem, should we consider increasing employment as a positive factor in assessing the appropriateness of a technology? What is the human cost of using technologies that employ fewer and fewer people? When is it appropriate to use machines rather than people? What about when you can get a machine to do a job that would be unhealthy, dangerous or just extremely unpleasant?

Do we really need all these different food items?

Fig. 8.3 *Average consumption of energy (in gigajoules)*

6 The use of appropriate technology should result in self-respect and increased self-reliance

The impact of a technology has to be considered from all aspects of its effect on people (individuals and communities), and its effect on people's self-esteem is part of this.

7 It should use renewable sources of energy wherever possible, and be economical on the use of non-renewable sources.

Will we have any future at all if we do not do this?

In the industrialised West we are consuming energy, mainly from non-renewable sources, on a far larger scale than in the Majority World. The average British person consumes 150 gigajoules per year, whereas the average person in India consumes 8, in Nigeria 5 and in Tanzania only 1.

8 The technology should fit in with the local social and cultural environment

Technologies will not work successfully in any setting unless they 'fit in'. But have we become very adaptable (too adaptable?) in the Minority World? If so, what has made us so adaptable? In what circumstances do people accept change?

Is there just one solution?

If we apply these criteria to technological choices here, it does not mean that there is one set technological solution to a particular problem, wherever it is. It does not mean that all solutions will be 'high tech' or all 'low tech', all large-scale or all small-scale. The solutions will differ in different contexts within Africa, just as they will within Britain.

If you want to produce electricity fairly cheaply, you have some money, live in a hilly area and have access to a regularly running stream which flows for a good distance on your land, then the most efficient solution for you is exactly the same, whether you are in Peru, Wales, Nepal or the Yorkshire Dales. You will use a Pelton wheel turbine with some sophisticated electronics to ensure that you can make best use of the electricity produced.

If you want to build an energy-efficient house, then your solution is going to be rather different in coastal Peru from coastal Wales. An appropriate technology may well involve 'high tech' solutions, particularly in the form of electronic controls. The strength of always starting designing from a context is that the solutions are sure to be different because no two contexts are the same. But certain criteria can be applied to every context. All contexts that pupils work with need to be set within the wider one of building a sustainable future for the whole population of the world. This should become so much a part of pupils' thinking that they include it in their analysis without prompting. This approach can be developed by questioning assumptions and decisions at all stages of the appraisal of designs. One could ask, for example:

- What is it for?
- Why produce or use it?
- Does anyone need it?
- Will it improve anyone's quality of life (users or producers)?
- Will it harm anyone's quality of life (users or producers)?

A small-scale hydro scheme generating electricity for a rural community in Nepal

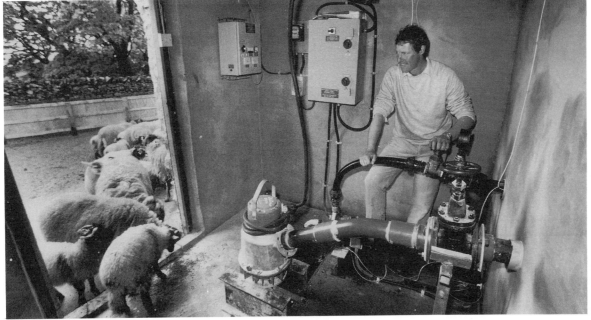

The Himalayan answer to a Yorkshireman's prayer! Micro-hydro power bringing reliable electricity to remote parts of Yorkshire

- Who would benefit from this product?
- Does it use irreplaceable and useful materials, and can this be justified?
- What better use might there be for these materials for which they should be conserved?
- What is the total energy used in its production (extracting the raw material, processing and transporting it and in disposing of waste when it has reached the end of its life)?
- What energy will be consumed in using it?
- What is the source of this energy?
- What pollution is being caused by constructing and using this?
- Does it promote physical health?

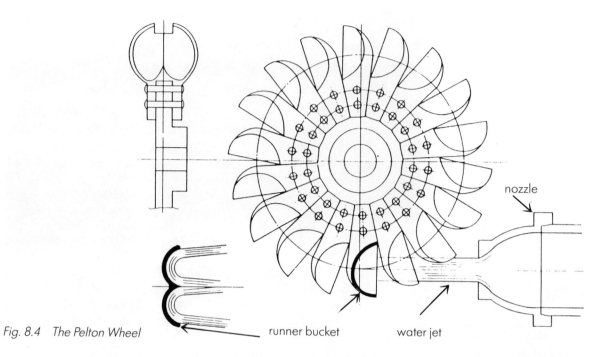

nozzle

Fig. 8.4   The Pelton Wheel

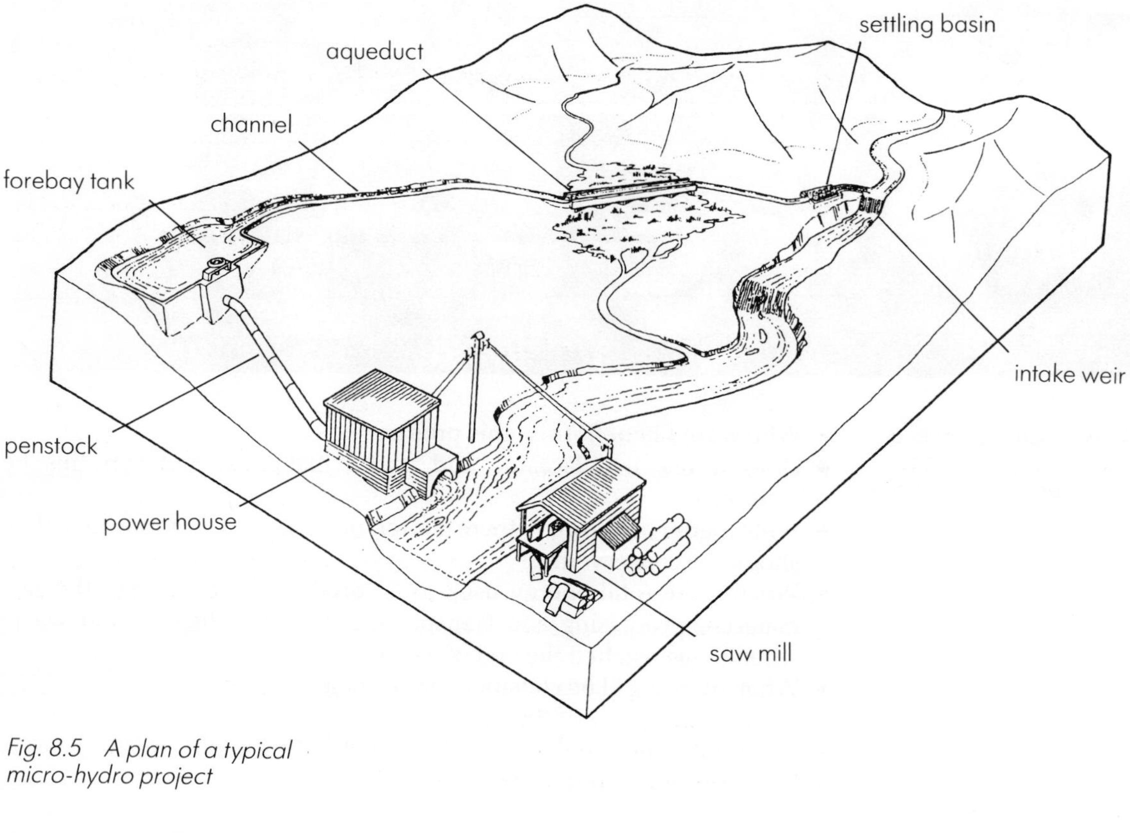

runner bucket   water jet

settling basin

aqueduct

channel

forebay tank

intake weir

penstock

power house

saw mill

Fig. 8.5   A plan of a typical micro-hydro project

Appraising and evaluating existing products is fun to do. If you look through a few magazines or catalogues you should find some good examples of products to consider. There is a fascinating book by Victor Papanek, called *Design for the Real World*, which describes some wonderful examples of human ingenuity put to total waste.

Needs and wants

There is a great deal of thought-provoking work to be done with pupils on the issue of the distinction between needs and wants. They may well not be aware of the concept of basic needs for food, clean water and shelter. If they look at the situation of a villager in Africa they will probably identify those as her basic needs. But in Britain most young people take those things so much for granted that they may well not see those as their own basic needs. Perhaps if one started with an image of a young person sleeping rough in Britain,* a young person of a similar age to that of the pupils, this might provide a thought-provoking situation. You could look at the needs of a homeless young person and then move on to looking at the pupils' own needs.

Start by brainstorming a list of needs with the whole class. Then divide them into groups (no more than four in a group), and get them to put the list in order of importance. Then brainstorm a list of wants with the whole class, and get them, in their small groups, to separate the needs from the wants. Then look at their lists as a class. Talk about the wants: does it make sense to prioritise them? (One would assume that they may be too individual to be able to compile an agreed order of priorities.)

If you can compile an agreed list of needs without an argument, it would be surprising. Do not avoid disagreements. A good argument is always an indication that there's some thinking going on, and if that is not happening you might almost as well be showing them all how to do secret mitred dovetail joints.

It is important that the pupils do not end up feeling that 'needs' are for the Majority World and that 'wants' are for the Minority World. What about fun in a world where one emphasises needs?

What about real life?

All this emphasis on needs may raise an image of a 'pure' future where we consume as few resources as possible, only use recyclable materials and renewable sources of energy, and do not consume any unnecessary products transported any great distance or anything unhealthy. What, no chocolate, coffee, tea, bananas or satsumas?

This takes us into the perhaps difficult and grey, but very important, area of compromise. Pleasure and fun are vital aspects of human life. Much enjoyment can obviously be had from things that do not cost resources (people in the Majority World are very good at partying and celebrating if they get half a chance), but pupils should also consider what use of resources we can justify for 'mere' fun.

* Julian Stapley makes the same sort of point in Chapter 12.

Above, celebration in Bangladesh; below, having fun in Zimbabwe – people can celebrate and enjoy themselves without using valuable resources

What is an appropriate method of transport?

Looking at transport is a good possible way of focusing what one means by an appropriate technology. It would be fairly easy to collect images of a variety of vehicles and transport systems (slides, advertisements and so on) so as to prompt thought. First, pupils could compile a list of what the various forms of transport are and what they need to know about them.

The forms could include taxi, bus, bicycle, boat, barge, tram, monorail, train, car (various sizes), plane, hovercraft, ferry and, indeed, legs! For example:

- What type of fuel does it consume when running?
- How much fuel does it consume per weight of one average person carried?
- How much weight and volume can it carry?
- How much and what sort of energy and raw materials did it consume to build?
- How many satisfying jobs were created to construct it?
- What is its projected lifetime?
- How easy is it for the user to repair?
- What speeds can it go at?
- What pollution does it cause?
- What risks are there for the user and other travellers they may meet?
- What is the environmental impact of various cars (for example, a large-engined one with lead-free petrol and catalytic converter and a small capacity engine with or without them)?
- How recyclable are its parts?
- What impact does the particular system have on the areas close to it?

Fig. 8.6 Different travellers and their situations

Pupils could compile an endless list of different travellers and their situations, but Figure 8.6 shows a few:

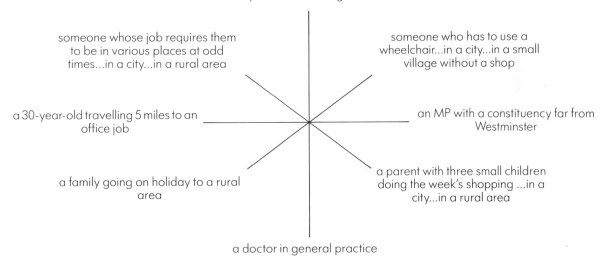

a 14-year-old travelling to school

someone whose job requires them to be in various places at odd times...in a city...in a rural area

someone who has to use a wheelchair...in a city...in a small village without a shop

a 30-year-old travelling 5 miles to an office job

an MP with a constituency far from Westminster

a family going on holiday to a rural area

a parent with three small children doing the week's shopping ...in a city...in a rural area

a doctor in general practice

Pupils could do this for different types of loads to be transported different distances. It is worth looking at how things were transported sixty years ago with far more use of trains and waterways. This leads on to looking at transport systems: road, rail and waterway networks.

Some thoughts: is the journey really necessary? Can the world afford foreign holidays in far-flung places for all of us every year? Who gets what out of tourism? Why is it that we have some cities that are very easy to drive across but where it is very difficult to walk or cycle from one place to another? Why has it become difficult and expensive to take bicycles on trains? Pupils could devise a campaign for persuading British Rail that their action in charging for bikes on Inter-City trains is environmentally unsound. This would prompt a whole lot of research and give opportunities for graphical communication and design and so on.

Pupils could design an appropriate transport system – for their own area, for a city, town or rural area, or even for the world! They could also design an artefact (some sort of device connected with bicycles or trains) or an environment (a railway station that is easy to use, pleasant to be in for a length of time and where you get healthy, reasonably priced fast food).

More appropriate contexts

One could look at many different things from this viewpoint. The questions suggested here are not a comprehensive list but just a few ideas. It is best if the pupils compile their own list through discussion.

- What is an appropriate fast food?
- When do you need fast food?
- How healthy should it be?
- Should it all be nutritionally balanced?
- What particular nutritional needs might there be in the situations where do you need fast food?
- How do you make it interesting?
- How far will the food have travelled, and with what transport costs?
- How do you package it appropriately?
- What's wrong with fish and chips wrapped in newspaper?
- What is appropriate packaging?
- What is it for?
- Does it need to be waterproof or airtight?
- Does it need to be protected from knocks?
- How many layers does it need?
- Will it be bio-degradable or recyclable?
- How much has existing packaging to do with advertising?
- Is the item to be packaged worth producing?
- What is an appropriate school lunch?

Pupils should not have much trouble compiling their own list of questions.

- What is appropriate leisure clothing for a young person?
- What can they or their parents afford?
- How long should it last?
- Who has produced the fabric, and where and what were they paid per hour?
- Does it use a natural fibre, and what are producers paid for the raw material?
- Does it use an oil-based fibre, and what more important uses might there be for that material?
- What is the role of advertising and profit-maximising in fashion?

Where are we going? An appropriate future?

The world around us and the changes happening to it provide ample scope for contexts for Design and Technology work. Appropriate contexts have to be real, not contrived. Pupils see through artificially contrived contexts very quickly. For a start, pupils of all ages are extremely interested in the vital issues of what the future of the world will be. They need little encouragement to see that it is more their future than their teachers'. They are able to take a broad view and consider all sorts of issues, and most of them are prepared to contemplate radical changes in life style. Try brainstorming what pupils (and parents, and teachers) might be prepared to 'do without'. An appropriate future is one which ensures some sort of sustainable future beyond the next fifty years for the whole population of the earth.

The school as context

The easiest way to start is with the school itself, looking at it within the context of its impact on the earth. The 'school' does not mean just the building but also the people, how they travel to it, what they eat inside it, and so on. Obviously, pupils need to understand the broader context. They need to know about the greenhouse effect, global warming, the ozone layer, acid rain, how long fossil and nuclear fuels will last and about the worldwide problems of poverty, powerlessness and exploitation.

Fig. 8.7   How much energy does the school use each day?

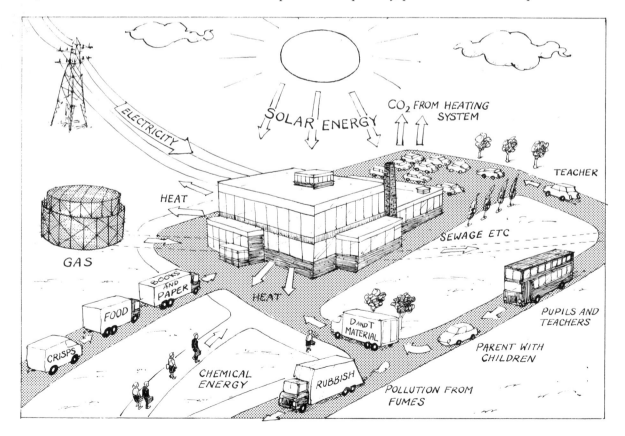

You may now be saying 'But I cannot cover all that in my one hour a week.' This subject is clearly very fertile ground for cross-curricular work, but you should also find that your pupils already know a lot about the issues involved and may have covered them thoroughly in some other area of the curriculum. Pupils can draw up a picture of the school. This could be a design or graphics project in itself, and could take all sorts of forms. It needs to show what the school consumes, what energy comes into it each day and what happens to that. This means looking at everything that comes into the school: chemical energy in pupils' bodies, paper for the photocopier, food for dinner and snacks, water from the mains, books (if you can still afford them), gas for heating and cooking, electricity for all sorts of things and, of course, all those raw materials used in Design and Technology. You should also add the transport energy involved in getting all those teachers and pupils to the school.

Then you have to look beyond the objects themselves to where they came from, how they were produced and what they cost. Economic awareness is vital, but it has to be an awareness not only of the here-and-now financial cost but also of the long-term, environmental and human costs of everything we consume. Clearly, from here various paths could be followed. Ideally, pupils should be identifying needs in areas in which they can work. This could be, for example, in energy production, saving energy, food production, cooking methods, recycling, cutting down pollution or creating wildlife habitats. They might decide that there were some personal or social needs that were not being met in the school, and that pupils' greatest unmet need was to have a place where they felt safe. They could design artefacts, develop systems or design ways of influencing other people's opinion or behaviour.

Rubbish

Pupils might design a recycling system for the school, with bins for sorting rubbish, garbage crushers and posters. In graphical communication they might devise graphics that discourage the production of rubbish; for instance, showing how the tuck shop might sell drinks in washable, re-usable containers rather than throw-away cans. They could design a bin that encourages people to use it by having some sort of visual or audio effect that is triggered when it is used. They would have to consider not only the waste of energy and resources in producing these throw-away items, but also the cost of disposing of them, the impact of landfill sites and the alternative possibilities of collecting methane or burning rubbish as a fuel. Whether to recycle and, if so, how to do so need to be considered. Pupils need to ask:

- What is the process used to recycle this material?
- How much energy does the process itself use?
- How much energy needs be used collecting and transporting the scrap? (This relates to its weight, the scale of the recycling process and how local it can be.)
- How does this compare with the energy used in extracting the raw material and processing it?

- How large are the estimated world resources of this material?
- Is there a better alternative in the first place for particular uses?

Energy

Most buildings in Britain are very inefficient in energy use, and your school is unlikely to be an exception. Pupils could look at the school's energy use and design improved systems or environments. This would involve analysing the rooms, corridors and so on in order to see when and how they are used and how much sunlight they receive, and therefore what level and type of heating and lighting they need. Pupils could evaluate artefacts such as various types of light bulb or draught-stripping systems, and design artefacts themselves. This could be a real-life example of economic awareness. The pupils could draw up accurately costed plans to submit to the governors for practical consideration. This could include: a lighting system that maximises the use of natural light and uses low-energy bulbs, an individual system of thermostats for every room to cope with different requirements, a draught-stripping and insulation plan, a scheme to encourage the users of the building to manage their energy use better (for example, by turning off lights or closing doors).

Very few schools would be in the sort of geographical position where one could suggest to the governors that it would be in their financial interests to install a wind or water generator. Schools do not tend to be built in very windy spots. However, since the electricity currently used by the school will be being generated some distance away, there is no reason why pupils should not experiment with wind and water generators and draw up theoretical schemes for wind farms many miles away. There is plenty of scope for investigation, experimentation and design with wind and water generators and the production of electricity from solar cells, as well as solar water and space heating systems.

Looking outwards

From the school one can move out to contexts within the community. The pupils' own homes is one area. They could look at their own energy use and the potential for recycling. These raise the issue of what is appropriate for individual action and what can really only be done by government or councils and what role pupils and parents might have in influencing policy. If pupils live in rented accommodation there is very little that it makes sense for them to do themselves in their homes to save energy, but they could lobby others to make changes. With recycling they need to look at the role of government and councils in facilitating the process. Then there is the whole practical problem of how you manage to sort and store various separate materials in a small kitchen. Can you keep compost on a balcony? If so, you could have the best-nourished pot plants in the block.

You could move on to looking at the community. What about public spaces and their use? What about a sense of community? How do you involve local people in decision making about their needs – designing with them rather than for them?

An ideal energy-efficient home

Designing an energy-efficient home makes an interesting activity. There are many issues to take into account, of which the first is orientation. In Britain we should build all our houses so that they can absorb as much solar energy as possible and retain it. This means that we should have lots of glass (double-glazed) on the south side and lots of insulation everywhere else. Conservatories are a good idea in energy-efficiency terms as well as for pleasure. This is an important point. We could build attractive houses that would be energy-efficient and also very pleasant to live in. This could allow pupils to bring their imagination into play, designing their ideal space to live in. To design such a house one would have to consider aesthetics, as well as inside and outside materials.

- Are they healthy to live with?
- Are they affordable by a family on an average income?
- Are they ecologically satisfactory to use (for example, not tropical hardwoods)?
- How much energy have they used to produce and transport?
- Who has produced them, where and at what cost to them?

Other factors to consider include insulation ergonomics, and area and arrangement of rooms, water heating, a back-up space heating system, and lighting systems, as well as the use of natural light and low-energy bulbs.

Design for fun

What about looking at celebration and fun as a context for design activity? Solutions should fulfil the criteria for an appropriate technology in terms of energy use, non-exploitation of others and other factors, but it is important to establish that having fun is a valid aim.

Special needs

Special physical or educational needs provide a wealth of appropriate contexts for designing. You can usually find a real context with a real need near at hand. Within the school itself pupils could look at how a pupil or a teacher in a wheelchair, or with some substantial difficulty in walking, would fare. In any class a few are sure to have a grandparent or neighbour who has difficulty undertaking some activity because of arthritis or failing eyesight. Young autistic children need toys which will attract them and encourage them to do things that they tend to need practice in, such as doing things in sequence. Here is great scope for moving toys with interesting effects. Children with cerebral palsy need all sorts of equipment to support them appropriately and encourage them to exercise in the right sort of way. Every such child is different, and what each needs is frequently not available.

Pupils are usually just as ready to design for the very obvious real needs of others as for the future of the planet, but ideally their work should be for someone who is real to them and fairly near at hand. It needs to be a context where, if they are successful, their solution will actually be used by someone. This of course points to the difficulties in examining technologies from 'other cultures', where you are saying the opposite: that young people are not designing *for* others, but they *are* attempting to understand, identify, and empathise *with* the needs of others.

Can you run out of appropriate contexts?

There seems little likelihood of this. We live in a very complex world, and 'appropriateness' as a concept can be universally applied *ad infinitum*.

When the pupils have absorbed the ideas of exploring real needs and of appropriateness, it seems very possible that they could start generating contexts themselves. It will be a sign of success if they start to scrutinise rigorously what you, the teacher, produce for them and reject some contexts on the grounds of inappropriateness. If they can say, 'We can't do this. It's wasteful of resources and wouldn't fulfil a real need for anyone,' then it was almost certainly a contrived context in the first place. It was probably an activity that had a context devised to lead to it.

I hope I have shown that the concept of an appropriate technology can be applied with gusto and success in a Minority World context, and that we would do well to apply the appropriate technology criteria to all our activities, inside and outside school.

Resources

Good Wood Manual and Good Wood Guide (1990) Friends of the Earth.
Lappe, Frances Moore 'Food for Thought' in *New Internationalist*.
Lappe, Frances Moore and Collins, Joseph (1980) *Food First*, Souvenir Press.
     (1988) *World Hunger: Twelve Myths*, Earthscan Publications.
New Internationalist, the monthly magazine, has many issues with relevance to the book. In particular, the November 1991 issue, 'Raw Food'. Address for subscriptions: 120-126 Lavender Avenue, Mitcham, Surrey CR4 3HP.
Norton, Prue and Symons, Gillian (1990) *Eco-school*, WWF.
Papanek, Victor (1985) *Design for the Real World*, Thames and Hudson.

Chapter 9 Two Case Studies

Martin Downie (1) and
Joan Samuel and Helen Abji (2)

1 Cooking up the Curriculum

So far this book has been rather theoretical; now, at last, we have a description of what actually did happen in a school in Cheshire. This was a very ambitious project, and broke much new ground, both in that particular school, and in Design and Technology-led cross-curricular work. What came out of it was a coherent educational experience, both for pupils and teachers, and teachers discovered, to their surprise, that departmental cooperation meant that the National Curriculum was easier to deliver. Martin Downie has written a fascinating account which captures much of the project's flavour.

In 1989, on a course for CDT specialists, I met a group of people involved in appropriate technology (from Intermediate Technology), who over a number of years had been adapting real life projects, from all parts of the world, for use in schools.

The projects dealt not only with the practical application of technology in helping to solve very real problems but also with the impact which that form of technology will have on the environment and people who use it. It seemed to me to provide an excellent starting point for a more holistic awareness within design and technology. One of the biggest problems I had found in the delivery of Design Technology was finding suitable project material that really did place technology within its proper social context, viewing it as a process, a human experience, rather than merely an understanding of gadgetry. Traditionally, many of these types of projects had been covered in the humanities curriculum area under Geography, where Development Education most naturally found a home. Now with the emergence of the Technology document, and a greater emphasis on designing for real needs, I felt these projects had a great deal to offer other curriculum areas.

Getting started

The project itself centres on the design implications of developing more fuel-efficient stoves in Sri Lanka, where growing fuel shortages through deforestation are having serious consequences for people and the environment. Having read the resource material on Sri Lanka and with support from other materials, I felt it could possibly provide the impetus for a major 'Technology across the curriculum' project. The complex issues raised by the project seemed to appeal naturally to a number of different curriculum areas.

Without too much effort I managed to get eight departments together for a meeting: Geography, Science, History, Art and Design, English, RE,

Making a fuel-efficient stove in Sri Lanka

Home Economics and CDT, to look at the 'possibility' of coordinating 'something' together. I had no clear idea as to what form that joint cooperation might take, and this proved to be my first of many mistakes!

We agreed to look at the materials I had provided and have a meeting at the start of the new academic year. What proved most interesting about that meeting was the various ways in which each subject area had interpreted the material. Each department had perceived 'their bit' and created a number of totally isolated and unrelated sets of activities, from making a paper elephant to comparative weather charts and installing gas central-heating systems. Working together was not going to be plain sailing: there was an obvious need to focus the activity, and after a great deal of discussion (sometimes heated) it was agreed that all departments involved should work towards one agreed aim – increasing their pupils' awareness of the role of appropriate technology in long-term and sustainable development of one particular country, Sri Lanka.

The task set was to construct a fuel-efficient stove to a standard which would enable the pupils to cook a Sri Lankan-style meal. To do this they would have to engage in the four elements of the design process:

1 the problem and its social context;
2 research and investigation;
3 designing and making;
4 evaluating.

It was also agreed that Year 9 would be used because Development Education was normally covered by humanities at that time.

Inevitably and appropriately, controversial issues had to be tackled: the whole issue of 'aid'; the complex issues of and reasons for poverty; and the opening up of traditionally closed subject boundaries. Probably the biggest (and most significant) mistake for me was to assume that colleagues, especially fellow Design Technology colleagues, would naturally welcome an opportunity to share their expertise, to discuss complex issues and further enrich their own understanding. The opportunity to explore and experiment with real issues was exciting and challenging, but the enthusiasm being created by the project frightened some teachers who compared it with what they were doing and saw the whole project as a threat to them. Where staff have a degree of authority, their 'fear' can be used to destabilise anything they perceive as a threat, and it is an issue that needs addressing if innovative work is to be encouraged in schools.

It became clear after the first few meetings that, as subject teachers, we all had our own curriculum language which was often used to cloud issues rather than clarify, and, more importantly, there was a limited understanding of what other colleagues actually did. Combined with fear about 'exposing' one's own expertise and knowledge with fellow colleagues, not to mention the inevitable clashes of personality found in any large organisation, much of the heated discussion was needed to clear the air and pave the way for greater trust and openness. An important part of my role as initiator of the project was not only to keep the project focused but also to encourage people to feel that they were all part of the delivery, and not merely contributing to or being used by the Technology element of the curriculum. Once we had begun to feel less threatened by

Fig. 9.1   A mapping chart for attainment targets

ACTIVITY \ AT	1	2	3	4
ART/CDT	Recognise that the likes and dislikes of users are important	Specify their intentions and needs by using simple drawings, models and plans	Demonstrate by their choice of tools that they understand the requirements for safety and accuracy. Apply knowledge of materials, components & processes to making	Justify the materials, techniques & components used, and indicate possible improvements
ENGLISH	Show judgement in the choice of information	Record the progress of their ideas, showing clarification and develpment		Appraise the outcome in terms of original needs and how it might be improved
GEOGRAPHY	Investigate contexts in a systematic way	Establish and check the availability of the resources required		Suggest improvements taking the user's view into account
HOME ECONOMICS	Recognise that the likes and dislikes of users are important		Identify sub-stages in their making and co-ordinate them into a simple plan	Understand that artefacts, systems and cultures have identifiable styles and characteristics
RE / HISTORY	Show judgement in the choice of information	Record the progress of their ideas, showing clarification and development		Understand that artefacts and systems reflect the circumstances of particular communities
SCIENCE	Use of both qualitative and quantitative data	Seek information from a range of sources and organise this to develop their ideas	Use knowledge of a range of materials to identify those most suited to their design	

one another and accepted the validity of all the contributions, we began to get to grips with the real issues of coordination and management.

In order to see how each aspect could be linked to the rest and how each department could constructively contribute, a few colleagues and I put aside one evening to map out the curriculum (see Figure 9.1). It is common knowledge that there is a great deal of overlap in the curriculum, but the extent of the repetition surprised many of us. A more coordinated curriculum freed departments to explore some issues in greater depth, because sections of their curriculum requirements could be better covered elsewhere. As we began to fulfil the requirements of the Design and Technology National Curriculum, we were also fulfilling large sections of a number of other curriculum attainment targets, and naturally vice-versa.

Delivery and resources

After consultation with relevant teachers, each department submitted their lesson plans, each taking into account the needs of other departments. In brief (see Figure 9.2), Geography delivered the background information and the criteria for appropriateness, the impact of aid and also looked at the causes of poverty. History looked at the historical role of imperialism and colonial rule and touched briefly on the present-day conflicts in the area. Home Economics studied the structure of family life in Sri Lanka, the role of women in society, the traditional diet and cooking methods, custom and practice. Art and CDT worked in close conjunction in the design and manufacture of clay and metal stoves, and examined how heat is generated and controlled. The production of a design folder pulled together all the relevant information gathered by the children. Science looked at the properties of materials, conduction, convection, insulation and also alternative forms of energy such as bio-gas systems, hydro-electric power schemes and so on. RE reviewed the impact religion has on people's lives and how that affects the decisions that you make when designing for other people.

It was made clear at all times that it would be wrong for teachers to give pupils the impression that they are 'solving the problems of the Third World': the real appreciation gained is designing and making with another part of the world in mind. This view was backed up and supported by the English department, who explored the use of language in our attitudes to race and other cultures.

The children worked individually in all subjects except Art, CDT and Home Economics, where the pupils were grouped into teams of five, each child with an identified role:

1 boss;
2 chief designer;
3 technical design engineer;
4 graphic and aesthetic specialist;
5 technician.

Some groups were all boys, others all girls, and some were mixed. The roles were discussed, and each pupil kept a record of their input into the group (see page 127, Evaluation).

National Curriculum Attainment Targets

SCIENCE

Special contribution to: making predictions, recording, observing, properties and use of materials, critical evaluation, combustion, biomass, recycling, environmental impact, forces, structures, human activities and impact, social/moral/spiritual and cultural contexts.

ATTAINMENT TARGETS

AT1 Exploration of science
Levels   3,4,5
AT5 Human influences
Levels   3,4,5,6,7
AT6 Types and uses of materials
Levels   3,4
AT7 Making new materials
Level   5
AT9 Earth and atmosphere
Level   3
AT 13 Energy
Levels   4,5,6,7
AT 17 The nature of science
Level   5

ENGLISH

Special contribution to: finding and selecting information, seeking answers, deduction, identifying key points, discussing complex issues, developing their own views, responding to questions, role-play, group discussion and presentation, variety of writing, communicating to others.

ATTAINMENT TARGETS

AT1 Speaking and listening
Levels   6,7,8
AT2 Reading
Levels   6d,7d,8c,8d
AT3 Writing
Levels   6a,6b,6d,7a, 7b,7d,7e,8b
AT4   Spelling
Levels   6,7

DESIGN and TECHNOLOGY

Special contribution to: material properties, mechanisms, forces, efficiency, testing, constraints, team work, value judgements, user consideration, design from other cultures, economic/social/environmental effects, evaluating artefacts, criteria for assessing, decision-making, discussion, reviewing needs and opportunities, resource availability.

ATTAINMENT TARGETS

AT1 Identifying needs and opportunities
Level 4   Statements   4a,c,d,f
Level 5   "   5b
Level 6   "   6b,c
Level 7   "   7a,c
AT2 Generating a design
Level 3   Statement   3c
Level 4   "   4c
Level 5   "   5b
Level 6   "   6c,d
Level 7   "   7b,c
AT3 Planning and making
Level 3   Statement   3b
Level 5   "   5b
Level 6   "   6b,d,e
Level 7   "   7a
AT4 Evaluating
Level 3   Statements   3a,b
Level 4   "   4c,d
Level 5   "   5a,b,c
Level 6   "   6a,b,e

HISTORY

Special contribution to: study of past non-European society (study unit C); history from a variety of perspectives - political, economic, technological etc; exploring links between history and other subjects; understanding how histories of different countries are linked; selecting and organizing historical information; asking questions about information.

ATTAINMENT TARGETS

AT1 Knowledge and understanding of history
Levels   3b,4a,4b,5c,6c
AT2 Interpretations of history
Levels   3,4,5,6
AT3 Use of historical sources
Levels   4,5

MATHEMATICS

Special contribution to: making and testing predictions, estimating measures, extracting information, recording, surveying opinions, understanding of numbers, percentages, structuring tasks.

ATTAINMENT TARGETS

AT1 Using and applying mathematics
Levels   3,4
AT3 Algebra
Levels   4,7
AT8 Measures
Levels   3,4,6
AT9 Using and applying mathematics
Levels   3,4,5,6,
AT12 Handling data
Level   4

GEOGRAPHY

Special contribution to: knowledge and understanding of places; similarities and differences between places; themes and issues in particular locations; settlements; economic activities; use and misuse of natural resources; quality and vulnerability of environments.

ATTAINMENT TARGETS

AT1 Geographical skills
Level   4e
AT2 Knowledge and understanding of places
Level   4e,5c,7g
AT3 Physical geography
Level   7d
AT4 Human geography
Level   6f
AT5 Environmental geography
Levels   3a,5b,6a,7b

Fig. 9.2   A D and T-led cross-curricular project focusing on energy efficient stoves. Taken from Stove Maker, Stove User, Teacher's Notes, *Intermediate Technology Publications, 1991*

The structure of the project was as important as the content: it was felt that we should attempt to deliver the project within the existing timetable structure, for two reasons: first, the less disruption it caused, the less hostility it would attract; second, because it was a 'traditional' type of school, if it couldn't work under those constraints it wouldn't have a long shelf life – it wouldn't be seen as 'proper' work. So it was decided that over a seven-week period all the departments would contribute when and where appropriate, using their allotted periods on the timetable. Some departments were more involved than others and contributed for only a few weeks, and some even continued the work into the following term. Often, when putting together a cross-curricular project, everyone feels they must be involved or are obliged to be so for the full duration, even when there is no valid contribution to be made.

The project was introduced to the whole year group on the first Monday afternoon with a short film and talk from the Education Officer at Intermediate Technology. In the final week there was one morning made available when the pupils could try out their stoves and attempt to cook a meal, weather permitting. This didn't require any changes in the timetable as so many subjects were involved and naturally taking place at the same time, so staff were always available and no special cover arrangements were required (see Figure 9.3).

The resourcing was initially a sensitive issue, but it was decided to break down the costs of what would normally be spent by each department on their own curriculum over a seven-week period. The costs of supplying all the needs for each pupil, paper, basic food and diverse material for stove manufacture (one stove per group), materials for testing and modelling, were also worked out, and a considerable saving was made by rationalising resources this way. The problem was to get people to pay up! For the purpose of this project I approached the

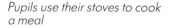

Pupils use their stoves to cook a meal

SUBJECT	Week 1	2	3	4	5	6	7	8+
ART CDT		X	X	X	X	X	X	
HISTORY	X	X	X					
GEOGRAPHY	X	X	X	X	X	X	X	X
HOME ECONOMICS		X	X	X	X	X	X	
ENGLISH					X	X	X	
SCIENCE		X	X	X			X	
RE	X	X	X	X	X	X	X	X

Fig. 9.3 The school timetable showing lessons devoted to the project

advisory staff and Cheshire TVEI which gave not only their moral support but also covered the cost of the project. The project was later used by them within the county to highlight good cross-curricular practice and innovative delivery of Technology.

Assessment and monitoring

It was felt all along that any form of assessment must be both easily manageable and easily understood by both child and parent. Each department constructed pyramids of attainment targets and levels (see Figure 9.4) in accordance with their own subject curriculum and programmes of study, and the pupils coloured them in as each level was reached and checked by the teacher. The pupil had to draw together in folder form all of the relevant inputs. The folder took the form of a GCSE design folder (the contents and depth of analysis of which often far outdid much fifth-year work). Where the children worked in groups, each child kept a diary to log his or her input over the week.

The ultimate test was of course to test the stoves and see how edible the food was, and whether there was a marked reduction in fuel consumption compared to the traditional open-fire cooking. For this purpose some simple scientific tests had to be devised.

Another aspect of this project was to see if we had affected some of the stereotypical attitudes to the so-called 'Third World'. This we attempted to do by devising a questionnaire that was answered before and after the project. Whether that worked or not was debatable, but as a result the Maths department felt that they could make a constructive contribution by analysing and presenting the results of the survey.

At the height of the project the pupil was concerned with the project in 80–90 per cent of his or her classes, covering all the areas traditionally treated in the various disciplines, but in a focused way. As a result, the quality and depth of understanding were there to see, the enthusiasm

Fig. 9.4   A pupil's assessment form for the project

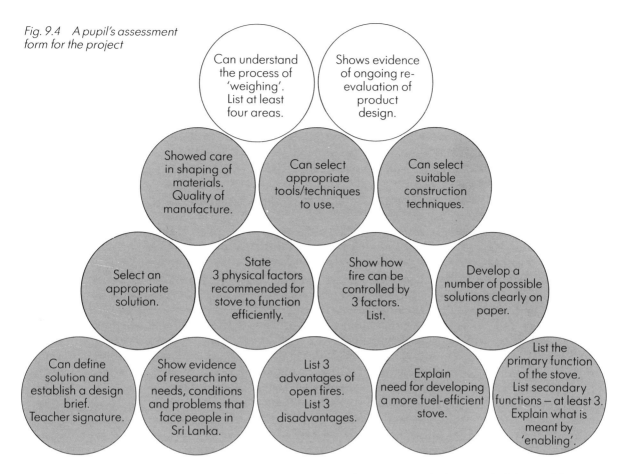

and energy were at times exhausting, and 'the proof of the pudding was in the eating'! As one pupil said to a representative from Intermediate Technology who attended on the final day, 'I know my stove looks like a lavatory, but it really works' (see photograph on page 128) – and it did, and the pupil was able to argue its advantages and disadvantages with disarming clarity.

Evaluation

At the end of the project it was felt that there was a need to gauge the feeling of the whole staff: twenty-five staff had been directly affected out of sixty, so a questionnaire was given out and forty-eight were returned. What follows are some of the benefits and problems identified by the staff:

1 *It provided an appropriate vehicle for the teaching of valid concepts in a variety of subject disciplines.*
'The practical element added spice to previously rather dry and abstract ideas.'

2 *The ending of the project was considered to be suitable.*
'To see the stoves working and be able to cook the food.'

One pupil remarked, 'I know it looks like a lavatory, but it really works.'

'Excellent! An enjoyable ending where the children enjoyed themselves and yet were able to show the concepts they have learned.'

3 There were opportunities for staff development, although these were not fully exploited.
'Greater contact and liaison amongst those involved would have reduced some behind-the-scenes tension.'
'I don't know, but I would like to come in on it more if it happens in future years.'

4 The contributions of individual staff were felt to be valued.
'Great – informal and liberal.'
'Reasonably happy, but felt a little in the dark at first.'

5 The majority of the staff involved were felt to be valued.
'Yes, good idea to cross subject barriers.'
'I would like to do it if I was better prepared.'

Lessons were learnt for a future project:

6 The introduction of the project to staff.
'Too much information for some staff to read – small amounts of input may be better.'
'Allow more time for staff to discuss how their individual subject objectives can be married together.'
'Staff need to be more aware of each other's input if the project is to bind together as a successful whole.'
'Too short notice but otherwise favourable, more preparation time, more integration with syllabus.

7 Possible alterations to the timetable were also commented upon.
'The project must be integrated with other work through the year – this was only five or six weeks.'

And pupil evaluation is also important. Figure 9.5 shows what one pupil thought.

This is my personal view of the project. It is clear that the National Curriculum supports the idea of a cross-curricular approach for Technology, but it is important that there should be clear principles on which that work should be based. The idea of technology should not be separated from the needs of people. *Technology is to do with people; it is for people.* As such it should not be taught to children without consideration being given to the social context within which the technology will operate. It should be based on problem solving and these problems should be real, not contrived.*

A child's learning experience is enhanced by personal involvement in a genuine issue. Technology is an area of experience, not a subject: it is an activity which crosses the curriculum, drawing on knowledge, understanding, skills and experience from across subject boundaries. Children's knowledge, understanding, skills and experiences should not be compartmentalised; there should be no artificial split of subjects into 'deliverers' and 'contributors' of technology: all should be contributors.

* This conclusion echoes Ann MacGarry's in Chapter 8.

During the Autumn Term we worked on a Sri Lanka project in almost every lesson. I think it was a good idea to use the same theme throughout as it helps us to understand the problems Sri Lanka faces today. For example, the geographical position and climate of Sri Lanka relates to its history, religion and people, and this in turn tells us about their diet and way of life.

The most clearly linked lessons were Science, Home Economics and CDT. In these subjects we learned about the shortage of electricity and gas supplies and how the Government is trying to encourage the Sri Lankans to use a new slow-burning stove. In Science we discussed how bio-gas generators work, while in CDT we designed and made a wood-burning stove. On this we prepared a typical Sri Lankan meal. I found these lessons interesting and fun, and they made me realise how much we take for granted – for example, the ease of food preparation and hygiene in this country. We also had the chance to prepare another Sri Lankan meal on an electric or gas cooker in Home Economics, which we took home for our families to try. The ingredients of these meals and the other information we had connected up with our Geography lessons, where we learned about crops, climate, natural vegetation, birth and death rates and distribution of the population.

In some subjects the six weeks were not long enough; for example, we did not cover all the tasks set in Geography; in History we did not have the time to learn about the rulers of the land, and in RE we only discussed the Buddhist religion. Maybe History and RE could have been more closely linked to tell us about the political problems of the eighties.

I would also have liked the chance to try out one of Sri Lanka's crafts, such as batik work, mask-making, etc.

I personally enjoyed the project but had the advantage of information, books, samples of crops, artwork and souvenirs, borrowed from friends who had spent several weeks there. The Sri Lankan Tourist Board, in London, also gave me interesting leaflets on the country.

Fig. 9.5 One satisfied customer at least! A pupil's evaluation of the project

A great many controversial issues were raised by the project and the debate will continue for a long time to come, but I believe pupils and staff emerged with a greater understanding of a far-off country and the role of appropriate technology and also an awareness of what is really possible within Design Technology. Hopefully, second time round it will be even better.

2 Design Technology for an Interdependent World: The Bedfordshire Approach

Here, in this second of two case studies, we have an absorbing account of what Alan Dyson advocated in Chapter 5: bringing about and managing curriculum change in a sustainable way, by involving teachers at all stages of the project development. The Bedfordshire multi-cultural services have taken a courageous and innovative stand in encouraging the introduction of a global perspective and multi-culturalism to Design and Technology teachers. These teachers truthfully voiced their misgivings at the outset, but since then have overcome their inhibitions. It is also an account of good cooperation between a range of agencies, with a specified goal for which to aim, and of sensitivity to the needs of the teachers involved.

Introduction: multi-cultural education and Technology

In this section we report on a development process which began in 1987. At that time, although the Centres had as part of their brief the promotion of multi-cultural and anti-racist approaches to all aspects of the school curriculum, there was, in practice, little development or expertise in what might be termed the 'hard' subject areas of the curriculum: Mathematics, Science and Technology. Most of the Centres' staff had themselves come from backgrounds in language or the arts. In addition to this, most schools in the county still perceived these 'hard' subject areas as culturally 'neutral' and therefore as having only tenuous links with multi-cultural education.

A major hindrance in developing our thinking about multi-cultural approaches to Design and Technology was the inescapable issue of aid to the Majority World. As Deputy Heads of Centres, one of whose functions was to promote positive images of cultures outside Europe, we had to ask how we could logically engage with an area of the curriculum which would seemingly point up contrasts in technological development and the need for aid, thus reinforcing existing negative stereotypes of people in the Majority World. While we could not deny the gross inequities that exist between North and South, we could, we thought, redress the imbalance in young people's perceptions of black people caused by media attention on disasters and charitable ventures. Our focus was always to emphasise the richness and diversity of language and culture and to promote anti-racist policies.

An important stepping-stone in the development of our understanding was the Centres' growing collection of curriculum resource material in the humanities. These expounded the concepts of the development education movement and world history. The former gave perspectives on global interdependence in terms of physical and economic resources in Geography, and the latter set British history in a wider context and gave a perspective on the effects of imperialist and colonialist systems on perceptions of both the past and the present. A number of these publications were, in fact, produced by aid agencies, and they were and

The title page from Design and Technology for an Interdependent World

are a real contribution to humanities education, but for the purposes of practical Design and Technology, while there was undoubtedly a connection, it was still unclear and our dilemma remained.

Beginnings

First contacts with Intermediate Technology (IT)

It was at this point that we came across the work of IT. In October 1987 Joan attended IT's annual supporters' meeting at the Central Hall, Westminster. Her motivation for doing so was mostly curiosity, prompted by a chance visit to the Intermediate Technology Bookshop. There was, too, the thought that she might find some resources for CDT. She was aware of worries in the black community about there being racist overtones in the philosophy of IT, so she was on her guard.

The subsequent report she wrote commented on the balance of concern in Schumacher's theories between the excesses of materialism in the West and the positive application of appropriate technology for the developing world. Her conclusion was that the IT approach was a properly anti-racist one.

It was apparent that great potential existed for using IT's work to introduce a multi-cultural and anti-racist dimension to CDT in schools, but the available materials at that time did not seem quite adequate for the task, though we could not articulate the reason at that early stage. It was also very clear that, even if the Centres possessed vast quantities of appropriate teaching materials, there was no demand for them from CDT teachers. The smart white folders of information about IT remained on the shelves, unused for another two or three years. Obviously, the need was to arouse the interest of teachers.

Beginning to raise awareness

Bedfordshire appointed a new County Inspector for Craft/Design/Technology in 1987. Martin Patterson came with an experience of working in Africa among his other many and valuable attributes. Martin already saw the need to introduce a global and multi-cultural dimension to the teaching of CDT, but he also, by now, knew his 'lads'; namely his predominantly male, practically minded, 'down-to-earth' group of teachers. He knew they were likely, initially at least, to resist change which they perceived as too political or abstract to be relevant to the practical business of 'making' in which they were primarily engaged. However, he and Joan planned a session for them, entitled 'Multiculture and Design (Technology)'.

With hindsight it is clear that the session was what it was intended to be – a start. It put issues of equal opportunities of both gender and race, as well as the global dimension, firmly on the CDT agenda. The presentations were arresting and thought-provoking, but the discussion showed how unfamiliar some of the concepts were to the participants and how much more remained to be done. The way forward in terms of equal opportunities issues was probably quite readily communicated. How cultural diversity and anti-racism, in line with the county policy on multi-cultural education, could be realistically incorporated into practical CDT activities remained something of a mystery. However, the session was important in that it became the beginning of the extended partnership in development – Multicultural Education Resources Centre (MERC), IT and CDT Education in Bedfordshire.

Developments

The National Curriculum

The arrival of the National Curriculum brought far-reaching changes which were to have important implications for what became Technology and its approach to multi-cultural education. First, the linking together of Art and Design, Food Studies, Textiles, Business Studies and CDT under the umbrella of Design Technology brought the CDT and Business Studies areas into contact with areas where there was a much more developed rationale and practice in the multi-cultural field – namely, Art and Design, Food Studies and Textiles. Second, there were numerous references in the common Statements of Attainment and Programmes of Study which gave opportunities for the exploration of cultural diversity. We still lacked any comprehensive National Curriculum cross-curricular guidelines on multi-cultural education, but the statutory orders for Design Technology gave us a head start, even if there was still more latitude than we would have liked.

It is greatly to the credit of Martin, and Kay Thurgesson, now adviser for Design Technology, that in setting up working groups to prepare different sections of the *National Curriculum Development Manual*, they did not lose sight of the on-going multi-cultural development process. It would have been so easy to have succumbed to the magnitude of the task in front of them and to have left it to the teachers to interpret the multi-cultural implications of the Statements of Attainment for themselves.

The Multicultural Working Group

Accordingly, a Multicultural Working Group was set up with the specific aims of exploring issues involved in bringing a multi-cultural anti-racist perspective to Design Technology and informing the planning for Key Stage Three National Curriculum training. It was immediately given status and seen to be taken seriously by the fact that the County Inspector decided to chair the group himself. In addition to the original partners, membership of the group was self-selected, but fortuitously represented a range of Bedfordshire schools, from urban high schools with a high proportion of pupils of ethnic minority group origin to mainly white rural upper schools, though there was a variability of attendance during its existence.

The first session

The first meeting of the Multicultural Working Group had the difficult task of defining terminology, setting long- and short-term goals, and working out strategies for integrating its work with that of the parallel National Curriculum task group which was in the process of designing training materials.

The then County Co-ordinator for Multicultural Education, Dr Kirit Modi, outlined the three major strands of the Bedfordshire Multicultural Policy, which places upon schools the responsibility for:

- meeting the specific needs of children of ethnic minority group origin;
- preparing all children for life in a multi-cultural community;
- combating racism.

This led to some very lively discussion, at the end of which agreement was reached about the necessity to produce a policy statement of Design Technology and Multicultural Education which would include practical guidelines, making it a useful working document rather than just a pious statement of intent.

The Multicultural Working Group did not wish to give marks out of ten to the National Curriculum task group for the extent to which they incorporated a multi-cultural dimension. They wanted, rather, to work in a supportive way to provide a 'cross-fertilisation mechanism'. Some of the pitfalls inherent in introducing aspects of cultural diversity into the curriculum, particularly the danger of reinforcing negative stereotypes of 'third world', 'poverty', 'deprivation' and so on, were aired at this meeting and would continue to be a major preoccupation throughout the project.

As to the development of a specifically multi-cultural worked example, it was felt that some input by a practising teacher who had already used some of the ideas the group had been discussing would be valuable at this point. It was decided to ask Colin Mulberg, who had helped develop and had used some of Intermediate Technology Group's materials.

Appropriate technology — Colin Mulberg

Colin's visit was a useful catalyst. He encouraged us to think about the social factors which provide the context for appropriate technology and our assumptions in tackling a topic, and went on to describe the work that he had been involved in developing with Intermediate Technology and the use of one of the modules with students. Teaching colleagues began to see the potential of these resources as models of good classroom practice and interesting motivators for children, but the wider issues of philosophy and attitude remained unresolved.

It is fair to say that there was some scepticism about the concepts of appropriate and intermediate technology within the group at the beginning. They were felt by some members to be too 'way out', 'radical', 'feminist' and so on. Certainly, they went much further than many teachers were prepared to go at that time, when some colleagues were willing to make a cursory nod in the direction of cultural diversity but not necessarily face up to what might become a challenge to many of their most deeply cherished attitudes – we might be asking them to examine themselves as well as what they were teaching.

One colleague, Bob Smith of Manshead Upper School, attempting to clarify the issues for himself, wrote about his concern that incorporating 'a multi-cultural dimension to our work should not mean that poor Design education takes place'. He wondered if there was a difference in content and style between the situation where 'ethnic minorities are the receivers' of Design education and where 'we are educating the white majority through Design to appreciate the culture and problems of other cultures than our own'. He felt that 'the context of a Design problem seems to be vital in introducing an ethnic dimension' and that wider issues such as language and terminology had to be considered. 'Students drawing on the culture of others to source their design ideas must not only observe and record the output of that culture but also understand the society and values which drive it.' He saw the need for work in Design Technology that was relevant to all children, but stressed that 'resources, both material, visual and human, are vital'.

Intermediate Technology (IT)

All these concerns were addressed at the next meeting, when we felt that we must move forward both on the philosophical and the practical level, and invited Catherine Budgett-Meakin to address the group. This was a very well-attended and lively meeting, in which participants were able to work through the idea of 'appropriate' technology and discuss IT's school projects in more detail. Catherine had also just been involved in the cross-curricular project in Cheshire described in the first part of this chapter, and presented an enthusiastic report back for what had clearly been an exciting piece of work both for the children and the teachers involved. The group decided to use one of IT's projects as a basis for its own work, and, after some discussion, chose the Bangladeshi textiles project for the following reasons:

1 within Bedfordshire, there is a sizeable population of Bangladeshi origin, which provides a rich resource;
2 the project could go some way towards breaking down negative stereotypes which might be damaging to the self-image of children of Bangladeshi origin born and brought up in this country;
3 it might serve to give a higher profile to Textiles, and provide a context which was more accessible to girls;
4 the Multicultural Education Resources Centres in both the north and south of the county already held a substantial collection of resources from Bangladesh.

Two examples from the Bangladesh Resource Collection: a jute door curtain (left) and details of a kantha (a hand-embroidered quilt) (right)

The group spent the next meeting viewing videos and slides, examining artefacts and discussing how best to go about presenting a view of Bangladesh which would be up-to-date and realistic without reinforcing stereotypes, and setting the context for children in a way which did not encourage the view that our Western society had the technological answers to all 'their' 'Third World' problems.

To demonstrate his commitment, Martin made £300 available to buy resources for a 'Design Technology Resource Collection' on Bangladesh. We visited the Commonwealth Institute and finally went to a supplier already well-known to us, Ruposhi Bangla in Tooting, where we found most of the resources and a great deal of help, experience and goodwill. From our point of view, this was a very exciting and enjoyable part of the project, and the constraints of the budget made us think very carefully about the appropriateness of the materials (see Appendix 9.3, page 146).

The *Development Manual*

By this time, the Multicultural Working Group had imperceptibly merged with the National Curriculum Task Group, both for practical reasons, such as the time factor, and because it was felt that the *Development Manual* must have a multi-cultural ethos built into it as well as setting out a specifically 'multi-cultural' module as a worked example.

The Bangladesh Resource Collection provided the stimulus at the next meeting of the group, which took the form of a working day to plan and write the section of the *Development Manual* which would focus on the multi-cultural perspective. We set up an exhibition and made a short presentation to the group, the focus of which was a video, 'Field of the Embroidered Quilt', which looks at the history, culture and life of

Finishing block-printed textiles in a women's cooperative in Bangladesh

Bangladesh from a Bangladeshi perspective, taking textiles and the involvement of women in their production as its starting point. We considered that the video should form the basis of initial in-service or preparatory work with teachers involved in the project in schools. It was considered important that the section should have a carefully worded introduction, setting out the principles informing the work, that the format of the module should be similar to others in the manual, and that it should contain practical suggestions and starting points for work which schools could develop.

The draft outline of the module was completed that day. At that stage, we were particularly happy with the introductory section. In addition to quoting from the County Multicultural Policy and seeking professional legitimation from the National Curriculum Design and Technology Statements of Attainment which referred to cultural diversity, it drew heavily on IT's pamphlet 'Strategies and Guidelines'. As a statement of anti-racist/multi-cultural aims, it was stronger and more explicit than anything included in other county training manuals for National Curriculum subject areas. Part of it is quoted in Figure 9.6.

Extracts from the Development Manual

It is important to be aware of some of the pitfalls that teachers who have not previously taught development issues might encounter. Primarily teachers are up against:

1 Their own prejudices.
2 The prejudices, racism and negative experiences of their pupils.
3 Our belief in the ethnic superiority of our society — both institutional and individual.
4 The negative stereotypes that people 'here' hold about people 'there'.

Assumptions to counter:
When using examples of Intermediate/Appropriate Technology it is important to be sensitive to any connotations of 'third rate' for 'third world'. Small-scale technologies can appear to be 'old-fashioned' and therefore 'second-best' or 'no-good'.
BIG is BEST
HIGH TECH is BETTER than LOW TECH
THE MORE IT COSTS, THE BETTER IT IS

Fig. 9.6 An extract from the IT pamphlet 'Strategies and Guidelines'

The title chosen for the module, 'Design and Technology for an Interdependent World', gave a global perspective to the work. We had been keen to build in the notion that the whole concept of the 'appropriateness' of technology was one which had relevance globally and not just in certain parts of the world, that technological decisions taken in one part of the world had repercussions elsewhere, and that our module on Bangladesh was not just a multi-cultural 'bolt-on' but could contribute to a wider learning experience about Britain's place in the world.

The worked example included suggestions for topics on 'Food for the Family in Bangladesh', 'Textiles' and 'Carrying Loads'. There was also a resources list and National Curriculum Planning Sheet showing which Statements of Attainment had been addressed.

This was very much a rushed piece of work, and perhaps none of those involved were satisfied with it. However, once again we felt that it was breaking new ground and that its significance lay in its existence rather than its level of perfection. Part of our problem lay in the fact that several members of the group, ourselves included, were not sufficiently aware of the structure of National Curriculum Design Technology; for example, the potential relationships between the four Attainment Targets and the nature of the specific skills to be developed. However, the next opportunity to clarify our thinking came during the Key Stage Three In-service Development Days.

Implementation

National Curriculum In-service Development Day

The full magnitude of the task presented by National Curriculum Design Technology became clear in the logistics of the first training exercise. But here again, in spite of a really packed programme, the multi-cultural dimension was included. There was a large display, in which the County Design Technology Resource collection on Bangladesh played a prominent part. Some Art and Design work based on an African theme from Manshead Upper School was also displayed. We were given a precious slot of twenty minutes in which to speak to multi-cultural and global issues. This time, we were able to focus on a series of Statements of Attainment drawn from Attainment Target One. We also drew attention to the relevant section in the *Development Manual* and the very helpful statement on Cultural Diversity in the Non-Statutory Guidance. We illustrated each statement of attainment with a colour slide which illustrated rural and urban life in Bangladesh. Here, at last, was a link between the humanities resources and Technology which we had been seeking; it seemed so obvious once we had made the connection.

AT1 4a	*Slide*
Starting with an unfamiliar situation, identify needs and opportunities for design and technological activity.	Jute workers in Bangladesh

The workers were engaged in a technological activity, taking up jute which had been soaking in the river. It was clear that this was an unfamiliar material, activity and context to many children.

AT1   4c   *Slides*

Recognise the points of view of others   Mrs Banerjee, wife of the headmaster
and consider what it is like to be in   of the village school (above)
another person's situation.   Mr Binod, local doctor/dispenser
                                                       (below)

Here was an opportunity to stress the importance of recognising the dignity of a
person from another culture by using their name and by presenting them positively
in an ordinary, day-to-day activity. Mrs Banerjee's point of view about food, as she
worked on the earth floor of her kitchen, would be different from the cultural norms
of the Minority World which are generally presented by schools. Mr Binod's view
of the delivery of health care would be strongly affected by his mode of transport
– his bicycle.

AT1   4f   *Slide*

Know that in the past and in other   A woman operating a dhekhi, a
cultures, people have used design   device used to separate grain from
and technology to solve familiar   chaff
problems in different ways.

This is a universal task, whatever the type of grain involved, and here is a simple,
effective method which is appropriate to the context.

AT1   5b   *Slides*

Recognise that economic, social,   A *kantha* – a hand-embroidered quilt
environmental and technological   traditionally made in Bangladesh
considerations and the preferences of
users are important in developing
opportunities.

Here the social, religious and historical background had determined the choice of
imagery used in the design of the quilt. The design was based on Buddhist, Muslim
and Hindu motifs.

AT1   6b   *Slides*

Explain how different cultures have   A bamboo baby protector
influenced design and technology,
both in the needs met and
opportunities identified.

The baby protector is an elegant and appropriate solution to a problem in a
context where a pram would be quite useless.

For some pupils who are of minority and ethnic group origin, the setting of Design and Technology tasks in Majority World countries will mean that AT1 3a is also appropriate:

AT1 3a	Slide
Starting with a familiar situation, use their knowledge and the results of investigation to identify needs and opportunities for design and technological activity.	Sujan outside a mosque in Dhaka

Sujan could be any secondary school pupil of Bangladeshi origin in Bedfordshire who is on an extended visit to his family's home. Such visits should be viewed positively as opportunities for Design Technology activity.

After this opportunity, we felt we had made considerable progress, but we knew that we had to do much more in order to fire the teachers' enthusiasm and to work out practical procedures for the classroom.

The Maryland Conference

The next step was to review developments to date and to plan for a Conference of Design Technology Coordinators. At the time, the thought of yet another mind-stretching effort was rather daunting. As generalists operating right across the curriculum and across age phases, we were both involved in a considerable range of other developmental work. At that stage, we had put Design Technology to the back of our minds. However, when we arrived at the next meeting, it was clear that Martin and Kay had been forging ahead. As well as the working group, we met there representatives of a number of other agencies: Mike Martin, of Intermediate Technology; Trixie Brabner, Head of the Schools Library Service; Diana Pollard from the Art and Design Unit; Alison MacFarlane, Crafts Officer from Eastern Arts. It seemed that we were to be engaged in more than a twenty-minute slot! In fact, the plan was that the session would occupy the bulk of the morning, lunch and the first part of the afternoon.

It turned out to be one of the most substantial multi-cultural In-Service Development (INSED) events, within a mainstream curriculum area, which either of us had experienced. The session was structured around a task which groups of teachers were asked to perform:
Plan a unit of work for Key Stage Three which

1 includes all the DT subject areas;
2 addresses National Curriculum Attainment Targets;
3 focuses on Bangladesh textiles;
4 avoids racist or sexist stereotypes.

The aim of the session was to allow the teachers an opportunity to immerse themselves in another culture which had quite definite links with our own multi-ethnic society in Bedfordshire; then, to get them to use their experience and expertise to devise practical tasks for the

Mehfil-E-Tar, Bedford Asian Women's Textile Group, who were involved with the Bedfordshire Design Technology Teachers' Conference

classroom. In order to provide a stimulating environment we created a Bangladesh Resources Fair, which transformed the whole of the lecture theatre at the Conference Centre and made a major impact on the seventy or so teachers who took part.

In addition to the agencies involved in the planning, all of which provided informative and arresting displays, there were some very important guests: Pervez Rahman, a Bangladeshi businessman involved in the textile trade, recommended by the Commonwealth Institute; Margaret Hakim, a local Headteacher, a regular visitor to Bangladesh through family connections, who loaned us part of her amazing collection of textiles and crafts from Bangladesh; Mehfil-E-Tar, which is a local Asian Women's embroidery and craft group set up through a Gulbenkian grant and now run by Bedfordshire Employment Training. Fifteen women from the group set up a workshop where they demonstrated textile crafts.

Although getting all these agencies together was a major organisational headache, the actual setting of the task for participants was perhaps the most difficult part of the whole exercise. We decided to use a pack of materials produced by Oxfam – 'Family Life in Bangladesh' – as a focus. This presented the context of a Bangladeshi village, Panishail. We used the visual material from the pack to create a display and adapted the background information. Groups of teachers were then asked to prepare a proposal for a term's work based on a particular scenario:

Let us imagine that a textile craft co-operative has been established in Panishail. Its intention is to make use of the existing craft skills of the local people and to introduce new, appropriate technology methods to make the enterprise more profitable. The object is to produce hand-made textiles for sale in the capital, Dhaka, and for export.

The intention behind this scenario was to incorporate the spirit of the National Curriculum Design Technology – that is, a progressive, developmental approach – with an understanding of the traditional values and skills of a culture such as that of Bangladesh.

During a very busy morning, the teachers were asked to visit the Resources Fair, to study the displays, which included videos and slide shows, to interact with the various county resource teams and visitors. Their reward at the end of the morning was an excellent Bangladeshi-style lunch – a 'first' for the Conference Centre catering staff.

Finally, there were brief presentations by Mike Martin and Pervez Rahman. Mike explained the philosophy of IT, relating it to the task in hand, and Pervez made a moving case for the need to preserve the craft traditions in Bangladesh. The teachers presented summaries of their proposals and Martin wound up this part of the day by stating, 'Neither you nor your pupils can solve the problems of the Third World. Your task is to broaden your pupils' understanding of the world in which they live.'

The way forward: developing practical activities

The teachers' evaluations of the day and their proposals for pupils' work revealed that they appreciated the need initially to immerse themselves and their pupils fully in any new cultural context. Although they felt it was relatively easy to devise suitable activities for Art, Food Studies and Textile work, there was still a difficulty about CDT and Business Studies classes. Some of them had found the textiles theme to be too restrictive, though others did see possibilities for related work in, for example, resistant materials. As INSED planners, we hoped that, given time, the teachers would accustom themselves to the ideas more thoroughly and that practical proposals would eventually emerge. However, we felt that the concepts of multi-cultural anti-racist education and appropriate technology and their relationship to the National Curriculum Statements of Attainment regarding cultural diversity were not yet embedded in their thinking. The Working Group met to review and evaluate the Conference, thus continuing the process of clarifying and refining its aims and planning the way forward. We wanted to build on what we saw as positive achievements, and so it was decided to engage in a dissemination process which would include the production of a short video about the Conference and a display of photographs, text and resources for the East of England Show in 1991.

Bearing in mind the reservations teachers had expressed, we also decided to work on proposals for practical activities which might address the structure of the National Curriculum more fundamentally, including the relationship between the Attainment Targets, namely:

1 Identifying Needs and Opportunities;
2 Generating a Design Proposal;
3 Planning and Making;
4 Evaluating.

The break-through in our thinking occurred when it was suggested that in dealing with a thematic Design Technology project set in a Majority World context, it would be necessary to break it down into two sequential stages. In the first stage, AT3 (Planning and Making) and AT4 (Evaluating) would mainly be addressed. Students would be introduced to the geographical, historical and cultural context and would then engage in practical 'design-and-make' activities which explored the traditional technologies of the culture concerned. They would then be asked to evaluate their products by applying a set of appropriate criteria. This first stage would enable them later to address the more sophisticated concepts of appropriate technology and to tackle the difficult task of AT1, Identifying Needs and Opportunities, in an unfamiliar context.

Stage Two is perhaps more likely to be covered at Key Stage Four. This could begin with a significant input to explain appropriate/intermediate technology, both in general terms and in its application to Majority World development. Students would return to the now familiar context, and begin to identify needs and opportunities for technological development using the IT approach (AT1). They could then plan for the production of artefacts, systems or environments which offer technologically appropriate solutions, both as original ideas and by studying real-life examples from IT projects around the world (AT2).

The question of whether they actually make a practical product at this stage is open to discussion. See Appendix 9.1 for the guidance which will be added to the *County Development Manual*.

To date, several middle schools have conducted cross-curricular projects based on themes such as 'India', 'the Caribbean' and 'Water in Africa', with varying degrees of involvement of Design Technology. Two upper schools have decided to engage in full-scale thematic projects based on the new county guidelines, with the whole of year 9: one is set in Bangladesh, the other in Africa. We eagerly await the outcomes and evaluations of these projects and the further developments to which they will undoubtedly lead.

Conclusion

We are aware that up to now the main beneficiaries of this process have been the members of the working group, though the importance of this should not be minimised. In fact, if we view the process we have described as an exercise in the management of change, we believe it illustrates most forcibly the need for time to be allowed for the trainers as 'agents of change' to develop their own thinking. They need the space and latitude to experiment, and to explore, implement and evaluate in a cyclic pattern. We are grateful to acknowledge that we were given such an opportunity, and hope that the process will continue and evolve within the schools for many years to come.

Appendix 9.1: planning process for a unit of work at Key Stage Three/Four

1 Select *context*; for example, business/industry/community;
2 select *theme*; such as a continent, country, culture, region;
3 select a *topic*, for instance, a craft cooperative or a market;
4 select *resources*, both human and material, which will illuminate the theme and the topic;
5 design a joint introductory session for all pupils involved which will help them to understand the theme and the topic.
   NB: It will be important to avoid reinforcing stereotypical images and concepts. It is probably best to dissociate DT projects from charitable aid initiatives. The intention will be to focus on the diversity, ingenuity and distinctiveness of cultural practices and technology. Avoid crude assumptions about 'high/Western tech = best; low/Third World tech = worst'.

It is suggested that the project is designed in two stages:

Stage one: AT3 and AT4

Design Brief One (for example)
A craft cooperative has been established in the village to produce a range of products using traditional techniques. You are expected to acquire skills in as many traditional technologies as possible, but you should choose and develop at least one into a finished product (AT3, Statements of Attainment 4c, 5b).

Evaluation
Evaluate the group of finished products together as stock for sale by the cooperative. Select those items which will be more viable as generators of income for a particular market, such as the local villagers or city shops, for export (AT4, Statements of Attainment 4d,5b,5c).

Stage two: AT1 and AT2

At this stage there should be a significant input to explain the concept of appropriate technology both in general terms and in its application to Majority World development.

Design Brief Two (for example)
The craft cooperative has selected items which will generate income. Suggest ways of introducing appropriate technology methods which will make the enterprise more successful. Choose one or more from the following list of criteria as the basis for your suggestion:

(a) aesthetic quality
(b) energy efficiency
(c) fuel efficiency
(d) cost-effectiveness
(e) labour-intensiveness
(f) durability
(g) environmental friendliness

(AT1, Statements of Attainment 4a, 4c, 4d, 4f, 5b, 6b, 7c,8c; AT2, Statements of Attainment 4a, 4c, 5b):
   This could be continued to AT3 and 4 provided that the work is linked with study of real-life examples of the application of appropriate technology.

Appendix 9.2: Suggestions for practical activities related to Art and Design, Textiles, Food Studies and Craft Design Technology

A craft cooperative

CONTEXT: Business/Industry/Community

THEME: Bangladesh

TOPIC: A Craft Cooperative

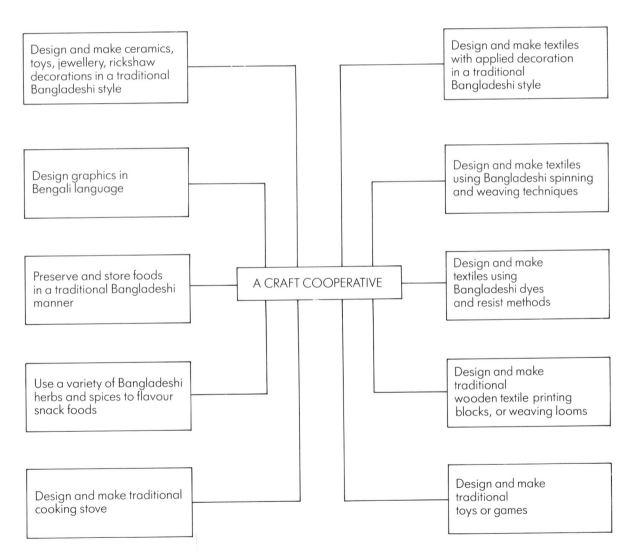

Design and make ceramics, toys, jewellery, rickshaw decorations in a traditional Bangladeshi style

Design graphics in Bengali language

Preserve and store foods in a traditional Bangladeshi manner

Use a variety of Bangladeshi herbs and spices to flavour snack foods

Design and make traditional cooking stove

A CRAFT COOPERATIVE

Design and make textiles with applied decoration in a traditional Bangladeshi style

Design and make textiles using Bangladeshi spinning and weaving techniques

Design and make textiles using Bangladeshi dyes and resist methods

Design and make traditional wooden textile printing blocks, or weaving looms

Design and make traditional toys or games

Information technology can, as a matter of policy be built into all these activities.

Appendix 9.3: Bangladesh Resource Collection for Design Technology

Textiles and artefacts	Supplier
Hand-woven raw silk	Ruposhi Bangla[1]
Nakshi Kantha	"
Jute table mat set	"
Jute door curtain	"
Floor mat *(pati)* – woven rushes	"
Bamboo fan	"
Recycled polystyrene fan	"
Kantha – white hand-embroidered	Shireen Akhbar

Books

Woven Air Exhibition Catalogue (1988) by Shireen Akhbar and Janis Jefferies	Bethnal Green Museum of Childhood[2]
Naksha, a Collection of Designs of Bangladesh (1981) by Sayyadar Ghuznavi	Ruposhi Bangla
Rangeen, Natural Dyes of Bangladesh (1987) by Sayyadar Ghuznari	
Rickshaws: Art and Industry, Basement Project, Young World Books	

Information on Bangladesh

Rickshaws	Basement Arts

Videos

'Field of the Embroidered Quilt'	Whitechapel Art Gallery
'A Visit to Bangladesh'	"

Slides

Slides on textile processes	Ruposhi Bangla
Slides on textile processes, Daily life, etc.	Intermediate Technology

Display materials

Woven Air postcards	Whitechapel Art Gallery
Greetings cards showing textile designs	Ruposhi Bangla
Bangladeshi Rickshaw postcards	British Museum Publications

[1] Ruposhi Bangla, 220 Tooting High Street, London SW17 0SG

[2] Bethnal Green Museum of Childhood, Cambridge Heath Road, London E2

Chapter 10 Technology and Theatre in Education

Michèle Young

Michèle Young gives us another tool with which to tackle our task. Theatre in Education (TIE) would not normally be associated with Technology education: here she describes some of the imaginative and innovative work that she and Passe-Partout have developed over the last five years. As she says, it all started with the Institution of Mechanical Engineers' Leonardo da Vinci Lecture Series in 1987, when Intermediate Technology was invited to present that season's lecture series. What came out of it was a 'theatrical presentation of IT's work', which received wide acclaim, and opened up the opportunities for TIE techniques to be used in Technology education. Michèle goes on to describe various possibilities which can be carried out in the classroom, which give an extraordinarily imaginative slant to classroom practice.

Part One: overview

Introduction

The juxtaposition of theatre and technology brings to mind the creativity and ingenuity of Leonardo da Vinci, a difficult act to follow. It is the novelty of this marriage of opposites which provides its strength.

Preamble

Children's Theatre is intended for entertainment; it sometimes also has a moral. Young People's Theatre refers to a form of amateur dramatics in which students can play a valuable part on the technical side. Without depreciating the value of these two forms of theatre, this chapter does not concern itself with either of them.

Background

Theatre-in-Education is twenty-five years old. Born out of the shift from teacher-centred transmission of knowledge to the child-centred approach of learning through experience, the idealism of the sixties created a climate in which it was possible to fuse Arts and Education. National and local funding of Theatre-in-Education initiatives and the emergence of courses at universities to train educators and actors in this sort of work indicate there was a social agenda.

Originally the work was quite often political in intention, where the desire was to bring about change. Existing texts were performed far less than work that was cooperatively devised, although this would sometimes incorporate the work of individual writers.

In the late seventies and early eighties the work spread from Britain to Canada, Australia, the United States and Europe. Meanwhile, in the United Kingdom, reductions in public expenditure into the nineties have meant that there have been shifts in the nature of productions.

A classic example of the use of Theatre-in-Education as a means of bringing to light the challenges of engineering in Majority World countries took place in 1987–88. The Institution of Mechanical Engineers' annual Leonardo da Vinci Lecture Series was in its thirty-fifth year, and the novel step was taken of commissioning a TIE production about the work of Intermediate Technology. This 75-minute presentation for, and with the participation of, fifth- and sixth-formers toured throughout the United Kingdom. Several follow-ons have since taken place in other countries. This innovative work illustrated a number of developments which were designed to encourage the exchange of technical information. In this way the challenges that face technologists and engineers can be shown from a perspective that is different from their own and is presented in a way that is perhaps more comprehensible, colourful, animated and accessible.

Aim

The aim of Theatre-in-Education is that the learning process should be memorable. Most conventionally this is approached by making the activity enjoyable: 'If you're laughing you're listening.'

Method

The techniques in these following examples have been chosen because elements of them can be replicated in the classroom. Some of the technologies referred to are drawn from those currently in use in certain countries of Asia, Africa and Latin America.

Drama used in this way becomes part of classroom activity. It does not take place on a stage, physically removed from the viewers. Pupil-led, it takes place amongst them. There are no boundaries between authors, actors and audience. The creative process of analysing the problem and then devising the format and building the script allows for as much of an insight into the technology as the final production. The method is inevitably cross-curricular: as well as the background knowledge involved it can contribute to developing analytical skills in science subjects, to developing linguistic skills through putting questions succinctly while maintaining accuracy, and to developing mathematical skills; for example, in devising fair scoring systems and methods of depicting team progress.

Part Two: body animation

Example One, the principle of the lever

The sunflower oil-seed press can be seen in Figure 10.1 and in Figure 10.2 (page 151) where it is represented by people through body animation. Getting students to think about the structure of a technology by attempting to build it with their bodies is very challenging. It is not hard to get them very quickly to put together an animated machine that resembles most technologies, but to get them to consider how to build one using their bodies as the component parts of a machine while

Extracting groundnut oil using a metal press, Malawi

illustrating its operation is much harder. The end result of a successful attempt can be most entertaining for those watching while the process for those involved has required careful structured planning, dealing with the mechanical problem of building from human bodies and team cooperation.

In the script below, which was prepared by the Kenyan group Pitia-Popote, you will see how the detailed instructions aim:

1 to make sure safety is observed at all times and that the participants are not embarrassed by what they are asked to do and are not uncomfortable for too long;
2 to ensure that the picture being built up is comprehensible and visible by the audience;
3 to use other members from the group chosen at random from the audience (this is desirable but not essential);
4 to illustrate clearly the principle of the lever.

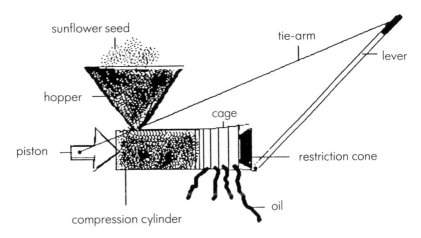

Fig. 10.1   The sunflower oil-seed press

ERASMUS   There is the diagrammatical representation of the machine.

BEN   And that is simple?!

ERASMUS   If you think it is somehow complicated we can recreate the machine here on stage.

BEN   Now you are talking.

ERASMUS   *(Calls his four team members)* Mueni *(she crouches on the floor)* represents the first lump of seed in our machine. Now at this point I need five volunteers from the audience to come over and demonstrate how the machine works.

BEN   Thank you for volunteering.... *(Five audience members are chosen and escorted to the stage.)*

ERASMUS   Welcome aboard, welcome. Can you all stand facing the audience? These two gentlemen will represent the compression cylinder *(They are guided across to stand facing each other, either side of Mueni, with their hands squarely on each other's shoulders)* in which the piston crushes the seed. Now our dear brother here represents the restriction cone *(He is positioned facing Mueni).*

BEN   The restriction cone?

ERASMUS   The restriction cone blocks the exit of the seed during the crushing process. The remaining two brothers will represent the cage of the machine through which the oil pours out *(They stand facing each other with the compression cylinder on one side and the restriction cone on the other).*

BEN   Already we have our first lump of seed. It has been crushed by O'Koth who is our piston *(O'Koth is solidly on all fours with his back ready briefly to support David, who is about to get in place)*, and now we need to pour in some more seed. David will represent the second lump of seed in the machine *(David, who is the lightest member, steps on O'Koth's back in between the compression cylinder, supporting his weight evenly on three points: their braced arms and O'Koth's back).*

ERASMUS   But how does the seed get into that position?

BEN   Usually there is a hopper through which we pour the sunflower seed *(raises arms high to show its shape).*

DAVID   *(the talking seed)* I can't go down because the piston is blocking me.

BEN   So I pull the piston back out so that the seed can descend *(David drops down to the ground).* To crush the seed I will push in the piston.

DAVID   Pathetic. Exert more pressure.

BEN   We have a problem. Do you have any suggestions as to how we can apply more pressure?

ERASMUS   If we could use a rope so we can pull the piston hard in.

BEN   OK. *(They weave the rope, which has the central point marked and easy to locate, under the piston's arms and through the compression cylinder's legs)* One, two, three, pull!

DAVID   Still not enough pressure. Try another method.

ERASMUS   How about a handle?

BEN   Precisely. Opiyo will represent the handle. *(Opiyo lies on his back with his head towards Mueni ready to do a shoulder stand, and the rope is tied to his knees.)*

ERASMUS   Now I'm going to pull the handle.

BEN   Still more pressure needed.

ERASMUS   How about a longer handle?

BEN   That indeed should give us that extra leverage. *(The rope is tied around Opiyo's ankles.)*

ERASMUS   Now I'm going to pull the handle so that it pushes the piston in this direction.

DAVID   Success, now I'm releasing oil. *(David's hands come shooting out between the cage's legs – the cage therefore must be wearing trousers for this to avoid embarrassment, and Mueni's head appears to the side of the restriction cone.)*

BEN   The hands here represent the cooking oil that we usually catch in buckets and purify by sieving. Over here the lovely head represents the seed-residue, that we can use as animal-feed or sell to other farmers. Now the operation of this machine is such that when the handle is up the piston goes out, so that the seed can descend; when the handle is down the piston crushes the seed. When it is up piston out, down crushes, up, out, down, crushes on and on.

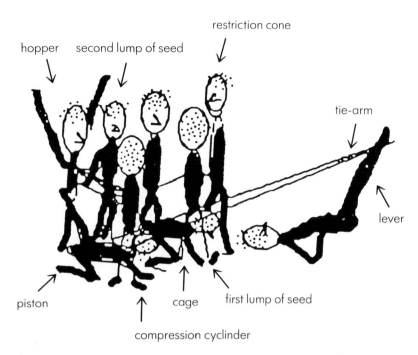

Fig. 10.2 Building a
sunflower oil-seed press using
bodies

Example Two, Stress = $\dfrac{\text{Mass}}{\text{Area}}$

It is estimated that by the year 2000, 100 million new homes will be required (adapted from UN Year of Shelter, 1987). The provision of cheap, durable building materials is an area beset with problems, more so when it is considered that the ideal manufacturers are small, local entrepreneurs. One of the difficulties is conforming to building regulations. It is essential to design quality-control tests which determine that the building materials are of the required strength. The means of testing must be scientifically exact and yet must not rely on expensive laboratory equipment.

An example of this is the quality-control test for Stabilised Soil Blocks. As the name suggests, this material for walls is made from soil and cement in the ratio 2:1. To check that the blocks have been pressed, cured and dried properly (they are sun-baked, which saves on fuel), the block is placed on two supports, 1 inch away from each side, and a mass of 90 kg is applied along the centre. The easiest way to do that is to get a man to stand on tiptoe in the centre on one foot, at the same time as lifting other blocks to bring his weight up to the 90 kg. This is quite possible and is entertaining to see – especially when the block does not measure up to the necessary strength!

Making stabilised soil blocks, Zimbabwe

A school built with stabilised soil blocks, Kenya

Research into building materials such as roofing tiles and blocks for building walls can also be done in class, and similar tests designed. The students' understanding of the principles involved in this method of testing could be aided by some questions of the kind given below.

QUESTION: Try snapping in two a piece of corn on the cob by holding it at both ends: it will break on the top side first. Why?

ANSWER: Materials are weaker in tension than in compression. When corn is being bent, the top side is in tension and the bottom side is in compression.

In the quality-control test we are measuring the block's strength in bending. Eventually, when enough load is added, the block will crack starting from the bottom, where it is in tension: not at the top where it is in compression.

QUESTION: When breaking a stick by holding it in your hands and stepping on the centre, is it easier to break it if you hold it near your foot or near the ends. Why?

ANSWER: The tension stress is bigger if you hold it at the ends
because the moment (moment = force × distance) is
bigger. As we saw with the blocks, they are placed as far
apart as possible so that a minimum force is needed to
break them. If they were closer together they would
require much more than 90 kg.

Part Three: ingenuity in design

Example Three, making one technology out of another

The students might put their minds to designing practical challenges for
a school quiz show. They prepare an 'assembling challenge' with pre-
selected materials for the teams to complete, pitting themselves against
the clock; for example, a hands-on attempt to make a simulated spinning
machine, where the aim is to get as much wool as possible spun onto a
bobbin in five minutes. Items required:

- a bike (best of all, a stationary exercise-bike);
- a broom handle with a doll's pram wheel fixed on it;
- two pieces of plastic tubing to make handles;
- a large bobbin (made from Sqezy bottles and a toilet roll);
- a ball of wool.

Fig. 10.3 Simulating the principle of a wind pump

A similar exercise simulating the principle of a wind-pump is illustrated
in Figure 10.3. Here air from the exhaust of a vacuum cleaner provides the
power to winch up a bucket using plastic cups and a bicycle.

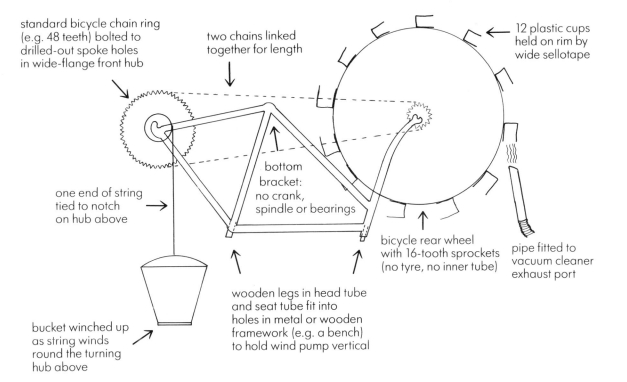

standard bicycle chain ring
(e.g. 48 teeth) bolted to
drilled-out spoke holes
in wide-flange front hub

two chains linked
together for length

12 plastic cups
held on rim by
wide sellotape

one end of string
tied to notch
on hub above

bottom
bracket:
no crank,
spindle or bearings

bicycle rear wheel
with 16-tooth sprockets
(no tyre, no inner tube)

pipe fitted to
vacuum cleaner
exhaust port

bucket winched up
as string winds
round the turning
hub above

wooden legs in head tube
and seat tube fit into
holes in metal or wooden
framework (e.g. a bench)
to hold wind pump vertical

Example Four, creativity in using simple materials

'Any fool can make for a dollar what only a genius can make for a dime.'
Henry Ford.

Or, as Schumacher said, 'Any third-rate engineer can make a complicated apparatus more complicated, but it takes a touch of genius to find one's way back to the basic principles.'

Making a levelling device for water-catchment projects in the Sahel is a cheap, practical challenge against the clock for one round of a game show.

There are two different levelling devices which could form the challenge. First, an A-frame plumb-line; second, for a more advanced group, a U-shaped water-level.

For the A-frame plumb-line each team will need:

- three sticks (two about 2 ft/60 cm long)
- four pieces of string (about 18 ins/45 cm long)
- one metal nut
- one marker pen.

For the U-shaped water-level each team will need:

- two sticks (about 2 ft/60 cm long)
- one see-through plastic tube (1 cm diameter, about 3 ft/90cm long)
- two pieces of string (about 1 ft/30 cm long)
- some water (in a little bottle)
- a marker pen.

The presenter needs a pile of books for the teams to test or prove their tools.

Here is the challenge.

In Africa, along the dry Sahel belt, there is a serious problem with soil erosion. In areas where the land is slightly undulating, low stone walls have traditionally been built to hold back some of the flash-flood water for just long enough to allow it to soak into the ground to germinate the local crop. But furrows and rivulets were constantly forming. The walls were vastly improved by the introduction of an affordable levelling device, which meant that the walls could follow the contours of the land, and thus help prevent the topsoil from being washed away.

The problem here is half solved for you. Build a tool which will enable you to find the horizontal; for instance, on this table [the presenter props three of the four table legs on piles of books of different heights]. Here are the components [hands each team their materials]. Are you ready, steady – go!

There are 30 points for the first team to finish, but it doesn't stop there because I will also be awarding points out of 100 for quality of construction.

The presenter ad-libs as the challenge progresses, then awards points to the first team to finish. The second team should be encouraged to finish within a minute or two of the first team – otherwise it might be advisable

Assessing levels for building water-catchment ponds, Turkana, Kenya

to tell them that they must stop with a part-finished levelling device. The presenter asks each team in turn to show how they use their level, then awards points for quality of construction.

This challenge can lead to some very amusing results, which make an ideal point of departure to discuss the various merits of the U-shaped water-level as opposed to the A-frame plumb-line in the situation described. Evidently, if sufficient materials were available to make very large models of the U-shaped water-levels, two people each holding one end could level large surface areas quite fast.

Part Four: game shows

The theory of game shows

The game-show format is the easiest dramatic structure to incorporate in classroom activities: it is also an effective way to create an exciting environment for learning.

Time: any class period can be adapted to a game show.

The versatility of the framework means that once you have identified the particular forms that you enjoy, they can be used repeatedly while varying the content. In leading or setting game shows 'practice makes perfect', or certainly practice leads to great innovations.

Game shows are based on closed questions, which are not suited to exploration or in-depth analysis at the time of the game show. They are particularly suited to introducing a new subject or to revision. Students need to be made aware that their competitive instincts can be used for the general enjoyment if they aim to involve the maximum number of participants.

The familiar game show routine comes easily to students who have had exposure to television. Designing and running a game show is both fun and demanding. The best material for your group is most likely to be that which you design for them or that which they are involved in preparing.

What is scripted for the teacher (or, preferably, for students who will be the presenters) need not be learnt or rehearsed, but if it is read out loud it must be read to the audience, not to the page. Your show might start something like this:

> Thank you for joining us on the show today. In Round One we are going to look at what we should consider when designing a new product or technology – that is: market, environment, needs and wants of customers, cost and job-creation; in other words – *is the price right?*

Using your own words is fine, but sufficient thought must go into the wording of the questions, especially when instructions are given. The desired answer(s) must be thought about ahead of time and clearly noted for the presenter, but a lot of thinking on your feet is required in order to refute incorrect answers accurately. Economy of words is essential to keep up the pace, but ambiguity must be avoided. Bear in mind the following:

- Focus clearly on the outcome you hope to reach.
- Keep wording simple.
- Devise questions which require short, quick answers.

After a quiz has been put together by a class and field-tested on its students it could provide a special event for the school.

What makes a game show effective?

- High standards of visual and oral communication stimulate and focus learning. Visual aids: slides, video footage, objects, pictures or practical problem solving all lend themselves to this approach.
- Changes of medium and pace revive attention and extend the amount which can be taken in at one time.
- A lively and witty approach to the subject breaks down initial reserve and guarantees a response.
- A chance for the audience to voice their questions and comments.

What makes an effective game show?

- A scoreboard is a useful prop. The whole issue of fair scoring and the method of recording the score is an interesting topic for discussion.
- Team signs are also helpful, as it is necessary to know who's who.
- Dividing the game into rounds allows for changes of team members, game format and subject.
- Music is optional and fun.
- Token prizes are an optional extra, or certificates can be drawn up.

Techniques of presentation

DO:
give clear explanations of all games.
repeat what teams have said for the audience to hear clearly.
give encouragement.
keep an accurate score.
reject wrong answers with a brief explanation.
have fun.
correct mistakes without drawing undue attention to them.

DON'T:
obstruct the teams' or the audience's view.
get side-tracked.
talk constantly.
forget to check all equipment before starting.

Fig. 10.4   Techniques of presentation

Pictionary

Pictionary is a 'game' providing key concepts in sketch form. Pictionary is like a dictionary, but it uses pictures instead of words. The aim is to be the first to guess what is being drawn.

The items to be drawn can be written down in advance and sprung on to the teams. Alternatively, the whole group can be divided into two, three or four teams, and can be given some time to choose quietly what it will give the other group to draw once Pictionary starts. Obviously, objects are easier than concepts; a bridge is easier to draw than metal fatigue (though it is surprising how quickly children can draw metal fatigue!).

Pairs

The game of Pairs can be used to take a look at roofing materials and their appropriateness for different climatic conditions and markets.

To assemble the Pairs chart you will need to have columns and rows marked by letters and numbers on one side of a piece of paper or card, and on the other side of each square a picture or word forming one part of the pairs described below. After the chart has been cut into squares, each one needs some form of adhesive on the picture side (Blu-Tack or Velcro) if you are to display them on a flat, vertical surface for all to see. Lay them out in their numerical and alphabetical row and column order.

In this example we have 20 cards: 10 words and 10 pictures – that is, 10 pairs. The pictures represent roofing materials and the words describe the environmental conditions in which they would be appropriate. The aim is to find the pairs. Each team takes turns to reveal two cards; if they are not a pair then they are turned back over again. If they are a pair, then that team gets 10 points.

The essence of the game of Pairs is memorising where each of the pictures or words lie hidden, but in this version, based on technology choice, the players must make a qualitative judgement of appropriateness based on economics, the availability of local materials and other factors such as the time scale or the local workforce. In many cases it might be argued there is no 'correct' answer, and often this dispute can generate a wealth of discussion about relevent and appropriate technologies.

Ten words describing roofing materials:

slate	clay
thatch	palm leaves
flat	concrete
wooden	corrugated iron
plastic	canvas

Ten pictures describing conditions where one of the above materials might be particularly appropriate:

Fig. 10.5   Playing pairs with roofing materials and environmental conditions

the English Lake District	savannah
desert	desert city
tropical coastal village	suburban housing
medical centre in	city slum
disaster relief camp	Himalayan mountainside
Brazilian rainforest	

Blockbusters

This game is a fast-moving, two-team challenge which can be worked on subject vocabulary or on a specific topic (200 words).

You will need to prepare a board like the one illustrated in Figure 10.6, on a black or white background. Most students will be familiar with this game, but otherwise the rules can be simply outlined as here.

PRESENTER: Now for Blockbusters. Team A is the red team and Team B is the green team. The aim is to make a chain across the board: red blocks for Team A and green for Team B. The way you win a block is if you are the first to answer my question. Each of the letters shown is the first letter of a missing word. Each time you guess a word correctly, a block of your colour covers that space and also you choose the letter for the next question. Team A is making a chain across the board and Team B is making a chain in a vertical direction, joining the top and the bottom of the board. Teams, you may choose to start wherever you like on the board.

PRESENTER: The team which answers this question gets to choose which letter they want first: C. What C force crushes the seed in a sunflower oil-seed press? (*Answer: compression.*)

PRESENTER: Well done, Team.... What letter would you like to start the game? I'll remind you, you are going from (*side to side* or *top to bottom*). Now teams, this question is for both of you. The first team to give me the correct answer gets the block and the chance to choose the next letter.

The set of all the questions and answers should be worded in the easiest way for the presenter – namely, with the letter clearly visible in the left-hand margin, then the question, always starting with the word 'What', followed by the letter and then the description. The answer is best written

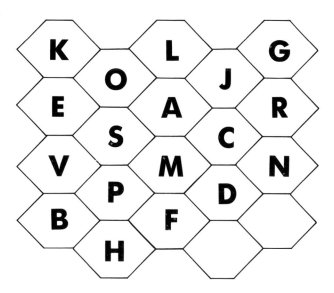

Fig. 10.6   A board for
'Blockbusters'

directly underneath. Ideally, the presenter will have two or three questions prepared for each letter just in case the teams are unable to answer the first. Colour-coding the questions according to their difficulty is quite helpful. Putting the questions in alphabetical order makes finding the letters easier, as does having them inserted in a flip-leaf photo album.

To get the groups to think up the questions to be asked of the other teams is a more demanding exercise than having them answer pre-prepared questions, and should only be tried with groups which have achieved some degree of familiarity with the workings of Blockbusters.

Part Five: conclusion

Comparisons and contrasts

Theatre-in-education shares with classroom teaching a sound basis of educational theory, in that it can be used:

- to provide experience in all the main fields of personal growth – namely, in intellectual, creative, emotional, spiritual and physical growth;
- to further the acquisition or extension of skills;
- to extend factual knowledge;
- to improve conceptual awareness and understanding.

As with other teaching methods, TIE programmes begin by defining the aims and objectives of the project, and then devise activities and experiences which enable them to be achieved. Reinforcement of learning and evaluation of the outcome are an important element which is built into the TIE method.

Some benefits of TIE

1 TIE programmes can be cross-curricular in method and content; for example, using dance to teach about science. This encourages the breakdown of barriers to learning and promotes lateral thought.

2 Presentations can use gestures, movement and non-verbal sound to communicate things which are not easily put into words, and introduce the audience to a fuller range of communicative powers. As the written word is increasingly supplemented by broadcast images, it becomes necessary to distinguish between fact and fiction in this new form.

3 The visual and aural content of TIE programmes makes them especially valuable for those who do not find it easy to learn through the written word. It also breaks down the usual stratification of the audience into the 'clever' and 'not so clever'.

4 The break from normal routine which the programmes represent has considerable impact. They jolt areas of the audience's thinking, cause them to look again at familiar things with more insight, and allow different pupils the opportunity to shine. Furthermore, the input comes through so many channels of sight, sound and sensation that normal inhibitions can be overcome and the audience's personal involvement stimulated very effectively. This is a valuable prelude to examination of their own thoughts and ideas.

5 The audience's powers of imagination are stimulated by TIE programmes, and, with imagination, can come enhanced empathetic awareness of other people. This is acknowledged to be very hard to achieve in everyday classroom learning, yet it is essential if people are to understand the position of other people in the world.

6 The audience's emotional responses can be stirred by a presentation in a safe and controlled environment and can be examined in a constructive way.

7 The considerable enjoyment which audiences experience in learning this way has useful on-going effects in other areas of learning.

8 The audience taking part in the programmes are as important as the actors in making the whole venture a success.

9 TIE is not something for which you can be too young or too old. It can embrace all groups regardless of race, sex or ability.

10 In a positive way, the relationship formed with the acting team is different from the one with the teacher. It demonstrates that learning can take place beyond the formal teaching context, and that new contacts and new viewpoints can be experiences to welcome.

Chapter 11 Creativity in Science and Technology

Mike Watts and Alan West

This chapter offers yet another practical opportunity for exploration. Using the vehicle of CREST (Creativity in Science and Technology), Mike Watts and Alan West explain the thinking behind the scheme, and draw attention to the importance of involving teachers at an early stage. They also describe some of the outcomes of the collaboration with the World Wide Fund for Nature (WWF). The concept, 'Think globally, act locally' is referred to, with examples of environmental education which have come via the CREST route.

Introduction

In the mid fifties Lord Ashby took the view that 'a student who can weave his technology into the fabric of society can claim to have a liberal education'. These days the saying might be rearranged: 'a society that can weave its technology into education can then claim to have a liberal student'. And by this we mean simply that curriculum development has reached the point when technology is seen to be part of the educational entitlement for all young people, and schools in the United Kingdom are now in the position of weaving technology into the curriculum on society's behalf (see *Technology and the National Curriculum*, 1990).

All this is fine by us, with one necessary provision: within the greater educational shifts taking place we would want to retain Ashby's original emphasis on the individual. It is the student who must see how technology can be woven into life, and key words in our lexicon as we work within science and technology are 'the ownership of learning'. The nature of our work is such that we want to encourage young people to take ownership of problems and solutions, during the course of which they offer their own creative talents to the process. This is an emphasis we have described elsewhere (Watts and West, 1991; Watts, 1991), though in this chapter we want to take a distinctive slant. Specifically, we want to focus on how youngsters might take ownership of problem solving triggered through the context of environmental issues, and to explore teachers' roles in this.

Stenhouse's classic (1975) on the curriculum makes the essential point that there can be no curriculum development without teachers' professional development – that the teachers are the school gatekeepers of the curriculum. If we want students to develop personal responsibility for their learning, and in particular their scientific and technological problem solving, we must approach this development through teachers in schools. And as the snake eats its tail, so it follows that teachers themselves must see the development of autonomous learning as an important element in their work – taking in the process responsibility for their own learning as they develop creativity in science and technology with students. Enter here the CREST project.

British Association for the Advancement of Science
The Standing Conference on Schools' Science and Technology

BRONZE AWARD

Presented to

For Creativity, Perseverance,
and the Application of Knowledge
in Science and Technology

SIR CLAUS MOSER K.C.B., C.B.E., F.B.A.
President of British Association

ROY ROBERTS C.B.E., F.Eng
Chairman of SCSST

An example of a CREST certificate

CREST

CREST stands for 'CREativity in Science and Technology', a project initially sponsored by the Department of Education and Science and supported jointly by the British Association for the Advancement of Science (BA) and the Standing Conference on Schools Science and Technology (SCSST). It is coordinated at a local level through Science and Technology Regional Organisations (SATROs) and by some local education authorities. It is basically an award scheme for rewarding youngsters' achievements in school-based project work: CREST's primary aim is to promote scientific and technological problem solving in the 11–18 age range. It builds on the Young Investigator's scheme, also supported by the BA, and which is targeted at the junior school age range, 8–12. Both schemes aim to complement normal school work and are non-competitive: youngsters gain recognition for their work through the scheme as a national project, and receive a Bronze, Silver or Gold Award.

Because CREST Awards are given to students who can demonstrate through their project work that they have met certain criteria based on the problem-solving process, the awards are criterion-referenced to these skills. The structure of the reporting process used to monitor the awards is designed to encourage critical evaluation of the processes and skills being used, together with enlightened scepticism on the part of the young person!

The CREST Award encourages (in its early phases) and requires (in its latter phases) students to identify and work on their own problems. These problems (opportunities/challenges) are identified through an 'active partnership' between the students or school, and the industrial/business community.

How, you might ask, can students be encouraged to identify their own problems? Much depends on the freedom to choose, and the flexibility of schools (and adaptability of teachers) in allowing students to pursue an investigative pathway, within the normal constraints that schooling places on any activity. Clearly, any linkage with issues affecting the environment is affected by these constraints; for example, constraint may arise through the sensitivities held by companies or other agencies about projecting an adverse image of themselves by declaring that they have environmental 'problems' to solve.* Approaching 'the good and the green' with any suggestion that they may have environmental problems at all may just result in students being shown the door rather than given access to an area of challenging work!

CREST places considerable emphasis on 'problem identification', 'negotiation' and links between the student or schools and a range of outside agencies. Students are encouraged to develop their own strategies and (within the safety of the laboratory or workshop) are allowed to experience the successes (and failures) associated with project management. Negotiation at regular intervals over the criteria provides powerful insight for the students. They are not told what to do but are helped to achieve what they want in pursuit of their project objectives. Active partnership is the key to this added value dimension of CREST, and we would also suggest that building effective partnerships with companies and agencies over a range of general activities may well be the best way to 'break into' the area of environment with them – once good relationships and mutual understanding have been achieved.

The quality of the students' experiences gained through a project are monitored, using a 'Profile of Problem-Solving Skills'. This asks them to provide evidence of activity from within their project which they consider demonstrates the CREST process criteria. A series of 'You Can' statements on a record card maintained by both students and teachers points the way towards iterative problem solving and successful project completion.

* This echoes the point made in Chapter 1 about 'Third World problem solving'.

The three stages of the CREST Award (Bronze, Silver and Gold), then, are addressed by an accumulating set of criteria – a ladder of achievement in problem-solving process skills. Therefore a major point of the project is that youngsters work on projects they feel they own. Individually or in groups, the project is one that they choose. Teachers may be influential at the point of choice – as may the CREST local organiser – but the emphasis is heavily weighted towards the youngster designing his or her own investigation. At the end, students are required to explain the development and outcomes of their projects to the outside agencies who have supported their work. The scheme works equally well for both individuals and team efforts. This is particularly so at the level of Silver and Gold where the scheme is looking for quite original (creative) work.

One of CREST's secondary aims is to promote closer working relationships between schools and engineering, industry and commerce in the outside world.

Environmental problem solving

One view is that environmental problems are a subset of all the possible problems in science and technology. Recent thinking, though, is that we can no longer sustain this perspective, and instead the converse is true: all scientific and technological problems have some environmental implication or dimension. That is, science and technology can no longer be isolated from wider environmental issues. In its 'Curriculum Guidance' series (No. 7), the National Curriculum Council (1990) observes that

> 'NCC has identified environmental education as one of the five cross-curricular themes about which it is giving initial guidance. The themes are interrelated and share many features. They have in common the capacity to promote discussion of values and beliefs, extend knowledge and understanding, encourage practical activities and decision making and further the interrelationship of the individual and the community'.

With this in mind, CREST has drawn together with the World Wide Fund for Nature (WWF) to develop and support problem solving stimulated by environmental issues within schools. Again, we have sought to work closely with teachers, supporting their school-based activities.

An INSET model

With WWF we have developed a model of in-service training for teachers and INSET providers, which relates to technological and environmental problem solving within primary, secondary and tertiary education. It is structured around a series of workshops, and draws upon a range of expertise – from both the teachers involved and many other AOTs (Adults Other than Teachers), including, on occasion, input from Intermediate Technology.

Fig. 11.1   The CREST record cards

CREST SILVER AWARD
PROFILE OF PROBLEM-SOLVING SKILLS
You and your teacher will meet on at least three occasions to talk
At these meetings your progress will be assessed using the state
goal or criterion must be met on at least one occasion during
this does not apply if a particular criterion is not relevant to you

Name
Project Title   Individ
CREST Project Number   Numbe

You can	1	2
Produce a workable idea/ideas in response to the problem identified.		
Find out information to help you with your investigation.		
Transfer science/technology ideas from familiar to new situations.		
Show originality in how you understand and interpret the problem set and in the ideas you have for solving it.		
Select/reject possible ways of doing the investigation and explain your reasons.		
Design a fair test and predict what results you expect.		
Attempt to control several interacting variables.		
Use your results to find out if they fit the idea being tested.		
Explain what your observations or results tell you in the light of what you are trying to find out.		
Attempt to cross check your results and explain the results that do not agree with the predictions you made.		
Alter and improve your investigation in the light of what you have found out.		
Work carefully and accurately.		
Record different ways of doing your investigation explaining the strengths and weaknesses of each approach.		
Record and explain why there may be more than one explanation for what you found		
Understand in what ways your experim prototype might not work in a differen situation i.e. prototype uncertainty/ac limitation of experiments		
Explain how far you got in your CRE project, what problems you found tried to overcome them.		

Signature of Student
Signature of Leader

CREST GOLD AWARD
PROFILE OF PROBLEM-SOLVING SKILLS
You and your teacher will meet on at least five occasions to talk about your project.
At these meetings your progress will be assessed using the statements below.Each
goal or criterion must be met on at least one occasion during the project. Of course,
this does not apply if a particular criterion is not relevant to your investigation.

Name
Project Title
Project Consultant(s) _____ Individual ☐ Team ☐ Tick one box

You can	\multicolumn{6}{c}{Number of times assessed}
	1	2	3	4	5	Optional
Produce a workable idea/ideas in response to the problem identified.						
Find out information to help you with your investigation. Liaise with your consultant.						
Transfer science/technology ideas from familiar to new situations.						
Show originality in how you understand and interpret the problem set and in the ideas you have for solving it.						
Select/reject possible ways of doing the investigation and explain your reasons.						
Design a fair test and predict what results you expect.						
Attempt to control several interacting variables.						
Use your results to find out if they fit the idea being tested.						
Explain what your observations or results tell you in the light of what you are trying to find out.						
Attempt to cross check your results and explain the results that do not agree with the predictions you made.						
Alter and improve your investigation in the light of what you have found out.						
Work carefully and accurately.						
Record different ways of doing your investigation explaining the strengths and weaknesses of each approach.						
Record and explain why there may be more than one explanation for what you found out.						
Understand in what ways your experiment or prototype might not work in a different situation i.e. prototype uncertainty/accuracy limitation of experiments						
Explain how far you got in your CREST Award project, what problems you found and how you tried to overcome them.						

Signature of Student _____ Date
Signature of Leader _____ Date

FINAL EVALUATION
The two sets of criteria are the goals
which each student is meant to achieve
during a project. It is the role of the
Evaluator to make sure the goals
relevant to the project have actually
been met.

The first set of criteria have been
used by the group leader to complete
the Profile of Problem-Solving Skills.The
profile will be used when the Final
Evaluation is made.

The Evaluator will decide whether **the
second set of criteria** have been
achieved at end of the project.

You should expect to
Negotiate successfully with a consultant
to mutually agree a project specification

Explain clearly your part in planning and
carrying out the project using your own
words.

Show that advice from a range of sources has
been taken into account in problem
generation, perception and reformulation.

Draw valid conclusions through an iterative
approach to problem-solving.

Apply a wide range of concepts and skills
precisely, using appropriate controls, to
solve a real-life problem which meets a
human need.

Explain how problems were overcome and
alternative solutions reached.

Produce a clear and concise record of the
project using technical language and style

Demonstrate the fitness for purpose of the
product of your work.

Suggest where appropriate, potential
industrial/social/commercial applications
of the project work.

Signature of Evaluator
Date

The workshops are based upon teachers having opportunities to solve problems themselves, exploring particular skills, contributing practical case-study knowledge from their own work, interacting with other educators in the same field, and establishing networks of industrial, commercial and environmental problem solving through talk and discussion, and through active problem solving at their own level.

The model is established upon several assumptions:

1 Problem solving incorporates skills of a high order which are an integral part of exploratory and innovatory work in schools – particularly as set out in the National Curriculum for Science and Technology.
2 Hands-on, practical activity is vital as a component of the training of individuals concerned both with the induction of teachers into the process of problem solving and of the trainers of the teachers.
3 The processes of 1 and 2 above cannot be accomplished easily in small pockets of time – time for thinking and space for working are important for the substantial re-think that may be necessary. Appreciation, development and dissemination of the skills of teaching problem solving cannot happen instantaneously.

The major objective behind the work has been to bring together teachers of science and of broad areas of technology to look at the potential for curriculum development and classroom practice for technological and environmental problem solving throughout the school. The workshop sessions have also attempted to develop closer working relationships between teachers and representatives from the business community, particularly in the areas of problem recognition and potential for project support.

The activities

Where possible, the workshops have taken place over a full day (sometimes two), and have usually been based in a school or college where there are laboratory and workshop facilities. Locations have ranged from Elgin to Edgbaston, Wycombe to the Wirral, Hull to Haslemere. Numbers have varied between twenty and fifty teachers, with a further complement of local advisers, industrialists, educationists, press, technical support and interested observers.

The heart of the process is for teachers to solve environmental problems at a level that tests their own ingenuity. It is from this that come all the discussions of classroom management, resource availability, school systems and structures, differentiation and progression, summative and formative assessment, time and people. The core theme, 'Think globally, act locally', has been at the heart of the INSET activity. These discussions, of course, do not happen just after the activity of problem solving – teachers are skilled pluralists of the first order. Even as the 'Why do that?', 'Who does what?' and 'Which bit goes where?' are decided, so the learning potential, instructional practicalities and institutional implications are debated. This is why time is so important: mixed groups of specialists bring such diverse experiences and differing opinions, so that the 'chemistry' of the group generates its own 'learning brew', and this process needs stimulus and opportunity in order to work.

The problems themselves are important too; they are the stimulus. The problems listed in Figure 11.2 have been used in different circumstances and have acted as stimulus for students too.

*Fig. 11.2   Examples of
environmental problems*

School ponds

Undertake an assay of the school pond to record the physical conditions (temperature gradients, light levels), the chemical composition (pH, mineral deposits) and the biological features it contains.

Community litter

Develop systems for reducing the levels of litter in the environment so that maximum recycling can take place with a minimum of social constraint.

Landfill

Explore the properties of a number of plastic containers with a view to their role within a major landfill site.

Photograph

Using ordinary household materials elevate a lightweight camera so that it can take an aerial photograph of the site.

Oil

Consider ways of transporting oil under difficult conditions (for example, at very low temperatures) through ecologically sensitive areas.

Packaging

Design a versatile packaging system for a range of woollen products so that the package uses the minimum of material, provides maximum protection and security for the goods, can be warehouse-stacked and uses recycled materials.

Birds

Develop a range of measures to attract a variety of birds back to an inner-city school area.

Reclaim

Design a scheme to reclaim a previously neglected site used for dumping, and turn it into a conservation area.

School

Put together the designs for an energy-efficient school – and look at the practicalities on-site of adapting an old school to make it more efficient.

Hide

Design and build a cheap hide/windbreak from which to observe wildlife.

Cans

Develop a practicable system to squash completely cans for recycling.

Greenhouse

Design an automatic greenhouse watering system to water plants while the owners are away on holiday.

Slurry

Produce a mechanism for injecting slurry into the soil to a depth of 10 cm in order to help prevent fertiliser run-off in heavy rains.

Soft Fruit

When soft fruits are frozen, ice crystals are formed. The crystal size is important, because if they are too large they rupture the cells and spoil the fruit. Investigate the relationship between freezing time and crystal size for different soft fruits.

Grape sugar

Grapes are sometimes partly frozen to concentrate the sugar in the must when they are crushed for fermentation. Investigate the relationship between the freeze temperature and sugar concentration so that the highest sugar level can be obtained by crushing.

Yeast

Explore mechanisms for reclaiming the yeast contained in the effluent from a large brewery as it is pumped into a river.

There are, of course, many more problems and activities suitable for experimental work by students – these are just a sample.

A system to squash cans for recycling

Our INSET mission

In all, the expectations, discussions and deliberations of the workshops have reinforced our commitment to a particular way of working in schools. We express this here in the form of a 'mission statement', as follows:

> It is a part of the curriculum entitlement of each youngster to have the opportunity to undertake individual and collaborative practical problem solving. Within this they must be encouraged to take ownership of problems and reach personal and group solutions through active experiential learning. Our task is to enable this to happen in as wide and varied a number of contexts as possible, and in particular to help focus upon technological solutions to environmental issues.

It is this kind of statement that underlines the activities of the workshops. There is one further feature worth describing: we instigated the writing of 'action memos'.

Although the workshops could meet some of our expectations, towards the end of the time together it became clear that many participants had developed numerous 'needs' as the discussions had progressed: they needed more time, more talk, more people, more resources and so on. Our strategy was then to ask them to direct their need in the form of 'action memos' to the people and institutions they felt might best have the resources to respond.

Outcomes

There has been a range of outcomes from our work so far – outcomes at different levels of specificity. First, teachers themselves have valued the workshops and have applauded the opportunity to learn directly both from 'experts' around, and from one another. They have developed new skills and approaches where, for instance, they have produced sophisticated 'green awareness' brochures for their school using airbrush design techniques; have met computer-aided-design (CAD) software for the first time and used it to explore models of systems; have encountered elements of the biochemistry of yeast through fermentation technology. They have certainly learned about industry's relationship with the environment and the fallibility of 'high-tech' solutions to all global problems.

They have learned, too, of different ways to manage their classrooms and organise their colleagues in school so as to find the best way for problem solving to take place. In some instances colleagues have shown the way and demonstrated how it is possible to encourage problem

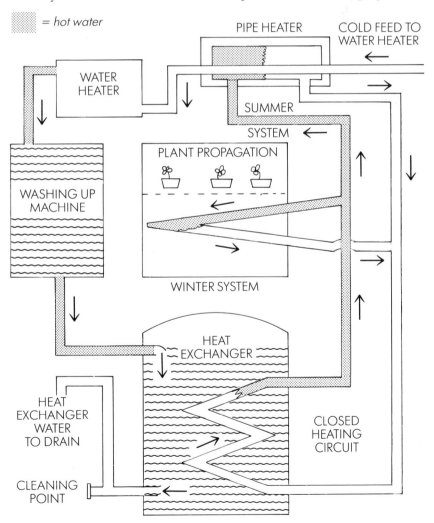

Fig. 11.3 Recycling heat from the school washing up machine

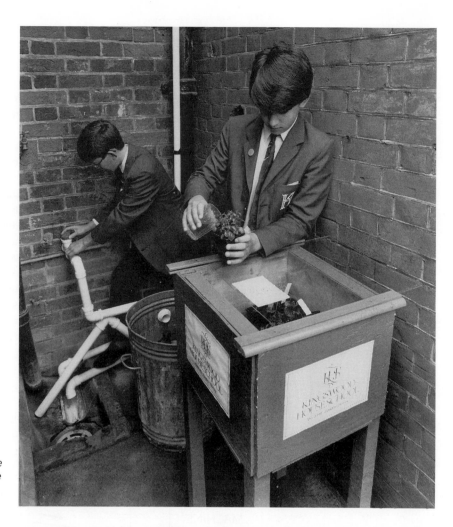

Recycling waste heat from the school dishwasher to help the school nature club grow their own seeds

solving in difficult situations; elsewhere they have studied cases of innovative work in different parts of the country.

Besides this, they have been introduced to global issues and the ways in which the activities they encourage at a local level can be seen to fit a context of 'thinking globally'. So, for instance, youngsters planning to turn their school into an 'inner-city oasis' within the East End of London have concentrated hard both at working within their own community and at the same time in developing a keen interest in the broader affairs of world-wide conservation. Their teacher has worked assiduously to encourage and protect plant and animal life within the school boundaries and, alongside them, to make contact with schools and organisations in other countries – initially, Hong Kong and Brazil.

At another level, the workshops have supported the growth of environmental problem solving in schools. The work with WWF has produced some excellent innovatory projects, many of which illustrate youngsters' ideas of technology appropriate to both the Minority and Majority Worlds. For instance, 11–year-olds at a Surrey school have looked for ways in which to improve their seed propagator. The drawback with a common cold-frame is that, although it provides some shelter for seedlings, it is still subject to sharp variation in temperature by day and night. The youngsters were aware, too, of issues of 'global warming' and the need to conserve fuel and recycle available energy. Their solution was to develop a heat converter using the hot water outflow of the school canteen's large dishwasher. They mended a big discarded metal dustbin which they filled from the outflow pipe of the dishwasher, and so provided a reservoir of hot water. The dishwasher is in heavy use, and the water emerging is very hot, and was previously pumped straight into the drains. In the bin they fitted an old radiator which acted as the heat converter, which in turn sent warm water through a system of pipes into their newly constructed propagator. They managed a temperature rise of about 10°C which proved highly effective in maintaining an appropriate climate for seedling growth. On the way, they had to solve technical problems of air locks in the water-flow system into and out of the bin; a filter system to prevent bits of debris from the dishwasher blocking pipes, and appropriate lagging for the bin. They experimented, too, with ways to monitor the effectiveness of the propagator to demonstrate the results.

At an important level, one further outcome from the workshops has been the confidence teachers have felt in raising the debate on environmental issues within their own schools. They have taken from the sessions the enthusiasm of others, the resources and materials of numerous organisations and concerns, and a network of like-minded supporters from different parts of the educational world.

Over the centuries it has never been particularly easy to involve people in specific ways of thinking and learning: attracting involvement in environmental issues in schools is still in its infancy. Our work has been geared to making some of that task easier for teachers as they undertake their daily work with youngsters in classrooms. Ideally, we want youngsters themselves to take ownership of the activities, since this spells the first step towards responsibility and involvement in depth.

As we noted earlier, teachers are the professional guardians of youngsters' learning and they have been our first port of call; to our way of thinking, a teacher won is another problem solved.

References

Curriculum Guidance 7: Environmental Education (1990) York: National Curriculum Council.

Stenhouse, L. (1975) *An Introduction to Curriculum Research and Development*, London: Heinneman Educational.

Technology in the National Curriculum (1990) Department of Education and Science, HMSO.

Watts, D.M. (1991) *The Science of Problem Solving*, London: Cassell Educational.

Watts, D.M. and West, A. (1991) 'Progress through problems, not recipes for disaster', *London Science Review*.

Chapter 12 Natural Technologists: The Contribution of Primary Schools

Julian Stapley

Although this book is intended primarily for a secondary teacher audience, it would be foolish to ignore and neglect the technology experience that children bring with them from primary schools. With emphasis being laid on 'cross-phase', Julian Stapley's chapter makes an important contribution. The reader will recognise many aspects in this chapter that have appeared elsewhere in the book. Young children, as Julian says, are natural technologists, and too often their learning, both inside and outside school, is not sufficiently acknowledged by the secondary traditional CDT teacher. Although for many primary schoolteachers, teaching 'technology' has appeared somewhat daunting, it is now becoming clear that much of what went on in primary classrooms constituted 'technology': the National Curriculum has mobilised and made respectable much classroom activity. There are also lessons to be learnt about cross-curricular work: it is, after all, the way primary schools work naturally. This can mean that the kinds of innovation suggested elsewhere in this book can be sustained more easily in the primary context.

Introduction

Children learn in many different ways, and their learning situations in school can, in many cases, be planned for in a sequenced way. Other school-based learning strategies can develop in a less formal context, and of course many learning situations can exist outside the school. What is fundamental to all good learning situations is the commitment of the learner: without this motivation the learning curve takes a sharp downturn.

Motivation is one of the 'keys' to opening the educational door, whatever the age of the learner. Motivation is perhaps the most essential ingredient of the educative process. Children, at the age of four, can arrive in our schools with a quite well-developed array of technological skills. Many nursery schools have groups of children 'playing' with water, floating and sinking objects, experimenting with pump-action toys, watching coloured water in one syringe travel along a tube to another syringe, realising that the more they press on the first syringe the quicker the other one shoots out. Other groups will be using real tools to manipulate wood – saws and hammers to create new shapes. The use of constructional 'toys' is another fertile area for very young technologists: their ability to 'design and make', based on what they know or imagine, can develop into a wide array of possibilities. However, if this 'play' is not structured or receives no 'teacher' input, a lot of this 'play' learning will not achieve its full potential.

It is perhaps through play that children are motivated to learn and so develop a sense of enquiry and curiosity. Play encourages children to be creative, to investigate, to begin to control what happens, and to draw their own conclusions.

Nursery children's work can lead them towards a variety of conclusions, and the teacher's role is clearly crucial. Their ability to value what the child brings with them, to provide an environment where learning can take place, their ability to enable, support and extend the children's learning is a vital one. Sharing as a partner in the learning process helps to ensure a broad and balanced curriculum.

This educative process was formalised by the Education Act of 1988 and the establishment of a National Curriculum for all pupils of compulsory school age. The Act provides for the Secretary of State to specify in relation to each core subject and other foundation subjects such Attainment Targets and programmes of study as are considered appropriate.

The identification of needs and opportunities by the pupils is, in my opinion, the most difficult of the five Technology Attainment Targets. It is of fundamental importance to this area of work, and AT1 gives relevance to much of the Technology work currently being developed in our schools.

If we take AT1, the *'Identification of Needs and Opportunities'*, and give it a global context, the possibilities are all too obvious, sometimes on a massive scale, and never with a quick, easy solution. At a basic level, the identification of need can be linked to fundamental human needs, and this is a major reason for the growth of technology. These fundamental human needs are numerous and varied, but, for many, satisfying the need for sustainable shelter, food and clean water are constant challenges.

Global issues

Many of these global issues have a rightful place within a school forum, and learners of all ages can appreciate, and respond, in varying ways, to many of these more globally identified needs and opportunities.

Shelter and food are both major human needs. Historically, and currently, there are many solutions to these problems: some reflect traditional locally available resources; some are a response to state-of-the-art materials. All reflect the inventiveness and ingenuity of mankind and womankind, irrespective of their socio-economic and cultural background.

At Key Stage One the problem of shelter and food can offer children a wide variety of learning situations. Some can be planned for, while others will arise naturally and, if time and resources allow, should be explored. Good primary practice must have imaginative planning at its core so that the teacher plans and creates situations where young and inexperienced learners can make informed choices about what they could do. It could be as simple as making a house, or a lunch set for teddy from cardboard boxes, or as complex as turning the home corner into a den area with different rooms. All the problems of access, light, supporting walls, roofing, and fitting out the rooms need to be addressed, as well as the social interaction of organising a party, role-play, drama and so on – the list is extensive.

Another 'list' a teacher must make is a conscious and methodical reference to National Curriculum Statements of Attainment. Primary teachers are very aware that activities like den building are cross-curricular, and children will deliver, in this case, Statements of Attainment from all the core subjects and various foundation subjects. This teacher-based assessment and evaluation is an essential part of classroom practice, and only by knowing what the child has experienced can the teacher plan for future learning.

The word *'evaluation'* appears in AT4 of Technology in the National Curriculum: *'Pupils should be able to develop, communicate and act upon an evaluation of the processes, products and effects of their design and technological activities.'*

The children's evaluation of their den will, at one level, allow them to change and modify their designs as they create. On a second level, there should be time for them to 'discuss with others', including their teacher, 'what they have done and how well they have done it'. Recognising and allowing time for this evaluation process is vital, as without spending time on AT4 any work undertaken cannot fulfil National Curriculum Technology requirements, nor indeed is it a complete educational experience without that evaluation.

At Key Stage Two as well as Key Stage One much work linking with shelter could arise from existing topic or project work, or perhaps from a science or story base. Careful planning is essential here as it could be very easy for the learning situation to be debased into a model-making experience.

Fig. 12.1   Case study: shelter

I recently visited a year 5 classroom where the children were involved with a topic studying North American Indians. The teacher's planning showed considerable resourcefulness and imagination, and the class seemed highly motivated and were producing work of high quality. To the rear of the classroom was a display area where there was a well-presented North American Indian village. Template-made fabric tents were mounted on cut dowling rods and stuck in plasticine to hold them up.

Next door was the other year 5 classroom, with children involved with the same topic – the North American Indians. At the rear of the classroom was a similar well-presented Indian village, although the tents were rather different. Some had the familiar shape but others had various extensions. One reason, I was told, was to allow for a large family; another was so that the family could have a bedroom and kitchen as well as a living room; a third example had an extension for a stable, to keep the family horses warm and dry. All these tents were mounted on twigs taken from a hedgerow – real wood chosen for a specific task. Some tents had attempts at waterproofing, and the nearby watering-can offered the chance for an objective and fair test for all tents. I remarked about the plastic horse in the display and was immediately told how important it was.

The class were aware of the semi-nomadic nature of the tribes and that the tents needed to be moved from time to time. Part of the evaluation of each tent was that it could be folded and lashed safely to the horse so that the tent could remain standing. This simple, yet objective evaluation can easily be assimilated by the child and can be related back to the original criteria concerning the needs of a semi-nomadic, horse-riding people.

Attainment Target 2 and 3 on their own can, and will, encourage some excellent model making, but to satisfy National Curriculum Technology all four attainment targets must be accessible within each experience or activity. Attainment Target 1 and 4 are therefore vital for the identification, development and assessment of Technology, and account for the real difference between what is Technology and what is model making.

These two teachers had planned and resourced their work together, yet the outcomes were rather different: this gives an indication of the flexibility available within Technology.

Another case study started its life as a project on 'the Weather', and also ended with 'Shelter'

Fig. 12.2 Case study: the weather and shelter

A group of year 6 children were studying the weather as a topic, and as part of the children's research they had to collect newspaper cuttings which highlighted the destructive force of the weather.

Within a week of starting the topic the news media was full of information about a cyclone's effect on the people of Bangladesh.

Very quickly a large amount of information and photographs were assembled, and the teacher sensed that her planned area of work might have to wait as interest and concern about the effect of the cyclone grew. From this situation it was possible for children to begin to identify needs and opportunities on such a massive scale that it was only through sensitive negotiation that the teacher was able to contain the situation. The children soon realised that whatever the solutions were, they needed to be affordable, sustainable, manageable, practical and acceptable to the local people. Various charitable organisations were approached for help, and the class chose to look at transport possibilities for flooded areas, and to try and develop some form of emergency shelters which could be parachuted to families whose homes had been washed away, and finally to organise a money-raising effort.

The emergency shelter idea was highlighted by a letter from a major charity stating that they had failed to solve this problem, and sent out rolls of polythene instead. How do you make a 'house' that can be packed flat, dropped by parachute, erected quickly and easily with instructions which could be used anywhere in the world – in other words, by illustration – and be robust enough to withstand prevailing conditions. Here was a quite complex design brief which year 6 children were attempting at their level. Quite obviously, it also has great potential at Key Stages Three and Four. However, on the development education front, care has to be taken that projects from an overseas context are not always 'disaster' ones, as that confirms the negative stereotypes that people hold.

The children started by investigating various items of packaging, looking at how 3D shapes were made – where were the folds? How could they fold flat? What were the mathematical nets of these containers? What might the net of a shelter look like? The children working on this project split into groups and set about designing and making their flat-pack shelter. They had access to card, fabric, string, paper fasteners and glue. Not only had they to design and make their shelter, they were also aware that as part of the evaluation, each was to be parachuted off the school roof and then constructed.

Another issue of wider significance was the fact that these shelters were only conceived for short-term use. What would be appropriate for long-term housing? What local materials are available? What expertise is at hand? What are the range of tools available? Are these 'Western-conceived solutions' always appropriate? Are the costs involved acceptable? Several children made a link between the card shelter they were making and the cardboard boxes regularly used by hundreds of British people as temporary shelters for the night! What are the moral, political and social issues here? Why do we have squatter camps and cardboard cities?*

The group organising the money-raising event had concentrated their efforts in two areas. The major event was to be a Saturday sale, a series of stalls set out in the school hall and run by the children. They set up ten stalls in all, including a crafts stall, to which the whole class contributed by making various items to sell. They also had a stall making and selling chapattis rather as Bangladeshi people might do. All the posters and other advertising was designed and made, as were the stall decorations; the group had organised a raffle, and various prizes were donated by local shops and businesses. The sale made a profit of £128. During the final week of term the group used some of their profits to finance a second chapatti sale, this time to other school friends. They had been able to experiment with various dough mixtures to find one that could be slit open and various fillings added. The fillings had been market-researched, and 215 chapattis were sold during a three-day period.

Outcomes

As a learning experience for children this project had a great deal to offer, but for the purposes of assessment children need to have tackled specific Attainment Targets. Work done should always relate to National Curriculum requirements.

English Attainment Targets are Speaking and Listening, Reading, Writing, Spelling and Handwriting. The majority of Statements of Attainment linked with each of these were delivered as the project progressed, and, given time and space, it would be possible to link each Statement of Attainment up to levels 5 and 6 with specific activities carried out by the children.

* This point links with that made by Ann MacGarry in Chapter 8.

Bangladesh is only in the news when there's a disaster. Here's a village eight weeks after serious floods

For Mathematics, Attainment Target 1; Using and Applying Mathematics, the Statements of Attainment up to level 6 were delivered in various ways. Similarly, numerous Statements of Attainment for the Attainment Target for Number were also dealt with, and the same applies to the Attainment Target for Measures, Shape and Space, and Handling Data.

The flat-pack shelter investigation also satisfied various Science Attainment Targets. AT1, Exploration of Science, states that activities should encourage the child's ability to plan, make hypotheses, predict, design and carry out investigations, interpret and communicate their results. Other ATs covered could include the Types and Uses of Materials; Explaining How Materials Behave; Forces and Energy.

The Attainment Targets for Technology in the examples cited above were all satisfied and the next task for the teacher was to look to the programmes of study as a planning aid for future projects. By careful record keeping the teacher was able to plot each child's progress, noting deficiencies which had to be put right during the next project.

Children beginning Key Stage Three should be able to do so with an array of knowledge, skills and attitudes which will allow them to progress smoothly into their secondary education.

If we follow the thread of the theme Shelter, Key Stage Three pupils should be able to exhibit a collection of research skills enabling them to draw on the technologies of other times and cultures as well as conventional technology practice. They should have improved access to industrial and commercial situations in order that they can become aware of production systems and be able to work in a cost-effective way.

Learning outside school

The children's learning environment is only partly contained within the school experience, and it is certainly more accountable within that school context. Children learn from a broad spectrum of experiences. It is unfortunate that many parents don't realise this and leave the whole learning experience to the school alone.

Outside the school environment children's play can be a powerful indicator of where their interests lie. Opportunities exist for choices to be made within extremely broad settings. Schools, by their very nature, can constrain and modify ideas, setting parameters that, after 4 p.m., do not exist.

Fig. 12.3   Case study: children's play

> Recently, a group of local primary schoolchildren have been involved with den building on a piece of waste land: an ideal opportunity of observing what children can do when school is finished and parents are not in view. The group played in a random fashion for several evenings, and then, as if consciously implementing Technology AT1 ('Identification of Needs and Opportunities'), they began to concentrate their activities in one area gathering together various materials. AT2 talks of 'Generating a Design Proposal'. Children explained their ideas to one another, using branches to show what could be built, pacing out a size, marking it out with stones, changing the shape to get more children inside. AT3 concerns itself with 'Planning and Making', and branches were duly rested across old pallets and thatching, and general infill followed. This first attempt was designed and made in one evening, and the group were able to evaluate one aspect of their efforts by all sitting inside.
>
> The following evening the group was larger, and re-designing ensued, and over a period of time a variety of changes took place. One door proved insufficient and others were built, and eventually after several more evenings of continuous activity they seemed satisfied.
>
> Here was a group of children, natural technologists, undertaking a task in a totally self-motivating and self-directed way: children being responsible for their own endeavours, independent learners involving themselves unconsciously in Technology and without knowing it delivering ATs from English, Mathematics and Science as well!

Another case study – 'Fiesta' – took strength from the multi-cultural riches of school and community.

Fig. 12.4   Case study: 'Fiesta'

> A fiesta is perhaps one of the most pervading examples of Caribbean culture: as a starting point for children at Key Stage Two it offers real opportunities to work in a cross-curricular way and yet deliver elements of all National Curricular subjects.
>
> A large, multi-ethnic middle school was looking for a project that would encourage greater links with the community and involve parents in an active way. The planning and resourcing of such an event is not to be taken lightly, and whole school planning is essential

if a quality learning experience is to be offered. The staff agreed that a fiesta should have all the potential they were looking for, and offered a perfect opportunity that could also involve both children, staff and parents.

For all the children to experience a fiesta at first hand would have been quite difficult, but the local community were able, in consultation with the staff, to stage all the elements of the fiesta so that children could appreciate them more easily.

The Monday morning of the fiesta week arrived, and the children were assembled in the school hall. The headteacher asked the children to sit quietly and close their eyes and to imagine they were going on a journey:

> Imagine the airport, it's cold and damp, you cross over to the plane and sit down, then you're moving, then you're flying, the countryside looks small, faraway, then there's sea – as far as you can see in each direction. The sun has appeared by now and the water takes on a blue tinge, dots of land appear, and at last the plane arrives. The door opens and you walk out, the heat, the brightness, the sounds, the fragrances, you have arrived!

A short general video of the Caribbean was followed by a local Jamaican-born man who cooked a specially prepared dish and offered tastings of several more. A story-teller followed, and two women brought their costumes last worn in the Notting Hill Carnival. The finale was a local steel band which entertained the school. Having been made aware of the resources available, the children were encouraged to make some choices. Since the children had a good working knowledge of Design and Technology, they were aware that they were looking for a need or opportunity to design and make. They were also very much aware that, to satisfy National Curriculum Design and Technology, making models of what they had seen was not sufficient; they knew that these starting points were just that and that their endeavours must exhibit a personal quality – the ownership must be theirs.

Using all their experiences, the group were asked to record as many needs and opportunities as possible, and from this wide selection of ideas the children had to focus down to specific work areas. One was resourced for food, one for music making and one for costume design and making, and the library was available for dance and drama activities. Each area was staffed by one teacher, various parents and people from the community, all of whom had been fully briefed, not only about the project, but also about their roles as adults: their intervention, use of open questioning and general presence. They were very much aware that they were to facilitate the children's ideas and not impose their own.

The group investigating food was looking, amongst many things, at relative cooking times, and used a traditional three-stone outdoor fire hearth, a purpose-made, clay, fuel-efficient stove, a household cooker and a microwave. They had various typical fruits and vegetables provided by a local Indian food store, and were encouraged under the watchful eye of the Jamaican cook to investigate both hot and cold main courses and desserts. These were then stored in the caretaker's deep-freeze.

The costume group were able to investigate at first hand the two carnival costumes, and to realise what use was made of wire and other strengthening materials, as well as that of fabric and paper.

With designs to hand, the group were able to start the making stage. Evaluation must be an integral part of any design-and-make activity, and can make you realise that your design is perhaps too tall and you can't make it stable.

As part of the planning the staff were looking for a balanced series of experiences for each child, and they had also planned for each child or group to be able to organise their own work time. The children were well used to working in this way, so they used this flexibility to its best advantage. Over the two days some children spent time in each area developing a series of experiences based on what was available. Other children concentrated more time in certain areas and developed work of greater depth and quality.

The fiesta days were split into two main parts. The morning session was an opportunity for children to refine their afternoon contribution and for the rest of the school to view what had been going on. It was also a valuable opportunity for all parents, governors, LEA officers and other interested parties to observe at first hand what a piece of extended experience work could be. Lunch-time had a special significance, as the usual school menu was supplemented by various more unusual dishes, mostly cooked by the kitchen staff, with contributions from the children, carefully supervised!

The afternoon session began with a series of stories and poems, some illustrated, some mimed, some with musical accompaniment. During the mid-afternoon break time and buffet the steel band was set up, and provided a suitable backing to the dance and costume display, which lasted till the end of the day.

Many visitors had stayed all day. Some had even joined in. No words can do justice to that spectacle.

Outcomes

The return to a more conventional school day was accepted, but the experience and excitement lingered on. The following Monday a child brought a head-dress he had made at home, entirely from materials which the child had found; others followed suit and within two weeks, twenty-three of the thirty-five pupils had brought in their head-dresses. They had designed and made them in their own time and were behaving as natural technologists, following through the four Attainment Targets for Design Technology capability. Children were identifying their own particular need and opportunity without a teacher in sight.

These highly motivated 'natural technologists' need, through a well-planned sequence of learning experiences, to be able to carry forward these attitudes, skills, concepts and knowledge not only through Key

Stages Three and Four but on into adult life. The necessity for youngsters to have the opportunity to design and make is crucial. What is also crucial is that their technological experiences do not solely reflect Minority World or 'rich-man's' technology: children must be able to draw from a wide variety of cultural backgrounds. These children must also be made aware that technology of any kind, at what ever level, must be appropriate for, and accepted by, the people who use it.

Conclusion

Technology must not become the ultimate arbiter of value, nor should we define progress in terms of technological specification. Can technology reduce complex human solutions to problems with controllable and quantifiable parameters, solved simply through rational analysis? Many primary children can appreciate the immensity and complexity of human problems; they can see massive food production and surplus set against widespread starvation; great advances in building science and materials technology set against a huge housing crisis. Younger children find it easy to empathise with others from other parts of the world, and to perceive injustice. As children mature, one of the greatest skills in dealing with technology is to be able to balance and negotiate social, cultural, technical, financial and political factors.

In order to ensure that a future that works, sensitivity is essential, and judgements about criteria, constraints and compromises can have their earliest roots in the primary classroom. Young children have a natural facility for learning, while we, as adults, are the custodians of knowledge and, as teachers, are the facilitators of learning. We must ensure that technology is seen to serve people, not the other way around or, as Schumacher said, we have to 'redirect technology so that it serves man instead of destroying him'.

Appendix 1 Intermediate Technology Educational Resources

IT education and source books for teachers

Intermediate Technology Introductory Slide Set

32 colour slides illustrating IT's contribution to meeting the basic needs for food, water, shelter, clothing and jobs; with notes and an introduction to the issues of development and aid.

Rural Blacksmith, Rural Businessman: Making and Selling Metal Goods in Malawi, by Val Rea and Mike Martin

Produced for Key Stage Three, this pack can be used within the Technology department, or as a cross-curricular project. The pack contains 30 colour slides in two sets: the first illustrates the context of the Malawian village from which students can identify needs and opportunities; the second illustrates indigenous solutions to the problem of air supply for the forge – invaluable for evaluation from 'other cultures'. An illustrated case study of the life of the rural blacksmith, together with an up-to-date country profile of Malawi, provides background material for students' investigations. For teachers, full information on the use of the pack is contained in the Teachers' Notes, and suggestions for tackling appropriate technology, global, and development issues are contained in Strategies and Guidelines.
1991, Intermediate Technology Publications.

Stove Maker, Stove User: Fuel-efficient Stoves in Sri Lanka, by Val Rea and Mike Martin

Stove Maker, Stove User is the second educational resource pack in ITDG's series 'Global Contexts for the UK National Curriculum', providing comparable information to *Rural Blacksmith, Rural Businessman*. The pack is particularly suitable to a cross-curricular approach. In this case the 'other culture' studied is Sri Lanka, and the issues rural people face in supplying energy for cooking. The case study of a family of potters describes not only the household's cooking requirements, but also their work as potters and potential stove makers.
1991, Intermediate Technology Publications.

Source books for teachers from Intermediate Technology Publications

Women and the Food Cycle: Case Studies and Technology Profiles UNIFEM

This collection of articles includes case studies of attempts to improve small-scale food processing, remembering that small is beautiful, but difficult.
86 pp., 1989, Intermediate Technology Publications.

Low-cost Printing for Development, by Jonathan Zeitlyn

For development and education workers, guidance on do-it-yourself printing methods which are simple and can be used with home-made equipment.
120 pp., 1988, Intermediate Technology Publications.

Educational Wooden Toys in Sri Lanka, by Tim Goodwin and Marjorie Wright

This case study on the work of ITDG and Sarvodaya, a Sri Lankan development charity, looks at the establishment of local production of pre-school educational toys.
56 pp., 1984, Intermediate Technology Publications.

Dyeing and Printing: A Handbook, by John Foulds

The text and line drawings describe chemical dyeing and printing techniques as they apply to small-scale operations, and how to plan for small-scale production.
48 pp., 1989, Intermediate Technology Publications.

Women and the Transport of Water, by Val Curtis

The haulage of water is one of the most arduous and time-consuming tasks of rural women, and this booklet looks at the scale of the problem in general and in particular in Kenya, suggesting ways in which improved methods of transport could help.
64 pp., 1986 Intermediate Technology Publications.

Small is Difficult: The Pangs and Successes of Small Boat Technology Transfer in South India, by Pierre Gillet

This case study describes the development, testing and production of fishing boats to meet the needs of local fishermen, now that their traditional craft are unobtainable.
Published privately, available from Intermediate Technology Publications.

The Copy Book: Copyright Free Illustrations for Development, selected and introduced by Bob Linney and Bruce Wilson

Over 100 pages of drawings drawn and donated by the British Association of Illustrators. Covers food, water, health, shelter and work. For those active in communication and education work.
102 pp., 1988, Intermediate Technology Publications.

Time to Listen: The Human Aspect in Development, by Laurence Taylor and Peter Jenkins

55 short case studies, based on experiences of VSO personnel, presenting situations that face rural and urban communities in developing countries. The reader is invited to consider possible solutions, but no definitive answers are presented. For prompting discussion at upper secondary level.
80 pp., 1989, Intermediate Technology Publications.

One Hundred Innovations for Development, edited by Sten Joste and Gillis Een

Technical problems require technical solutions, but however innovative the solution, it should also be simple, cheap, robust and easy to maintain. Lists 100 winning inventions in the 1st International Inventors Award competition, organised in Stockholm.
80 pp., 1988, Intermediate Technology Publications/ International Inventors Award.

Technology Choice: A Critique of the Appropriate Technology Movement, by Kelvin W. Willoughby

A comprehensive review and critique of the theory of appropriate technology which proposes a framework for integrating traditional economic development with its techniques – a sober look at the obstacles and an appreciation of its value.
350 pp., 1990, Intermediate Technology Publications.

Technology Transfer: Nine Case Studies, by Sosthenes Buatsi

Nine case studies in technology transfer which demonstrate the experience gained in different countries and different technologies.
82 pp., 1988, Intermediate Technology Publications.

Tinker, Tiller, Technical Change, edited by Matthew S. Gamser with Helen Appleton and Nicola Carter

Technical assistance fails to bring technical change because it fails to work with local innovation. This book raises awareness of people's indigenous innovation from actual cases in Asia, Africa and Latin America.
288 pp., 1990, Intermediate Technology Publications.

When Aid is No Help: How Projects Fail, and How They Could Succeed, by John Madeley with Mark Robinson, Paul Mosley, Rudra Prasas Dahal, Pramit Chaudhuri and Antony Ellman

Much development assistance from rich to poor countries has failed to get through to the poorest people, the ones who are most in need of assistance. Much official (government-to-government) aid has not even tried, but what about the aid projects that have genuinely tried to reach the poorest?
144 pp., 1991, Intermediate Technology Publications.

Where Credit is Due: Income-Generation Programmes for the Poor in Developing Countries, by Joe Remenyi

The author recommends the widespread support of credit-based income generation programmes as a cornerstone of a new poverty-orientated development strategy. The book argues that too little attention is devoted to poverty and the economic problems of the poor in the Third World and that what is needed is the courage to allow the poor to help themselves.
176 pp., 1991, Intermediate Technology Publications.

Energy Options: An Introduction to Small-scale Renewable Energy Technologies, edited and introduced by Drummond Hislop

Renewable energy can present a baffling array of options. While there can be no simple answer to the question of which energy technology is best in any given situation, the educator will find here useful information on which to base choices.
80 pp., 1991, Intermediate Technology Publications.

The AT Reader: The Theory and Practice of Appropriate Technology, Marilyn Carr

Bringing together a wealth of material, *The AT Reader* explains the idea, history, development and practical application of appropriate technologies through a wide range of contexts, and short case studies.
488 pp., 1985, Intermediate Technology Publications.

The Tech and Tools Book: A Guide to the Technologies Women are Using Worldwide, edited by J. Sandler and R. Sandhu

A research manual of appropriate technologies used throughout the world in women's projects. A useful book for teachers trying to get to grips with the appropriate technology, as it is well illustrated and laid out in such a way that pupils too would find it useful.
200 pp., 1986, Intermediate Technology Publications.

IT Publications has a catalogue, 'Books by Post', through which all these books can be bought. Alternatively, the bookshop at 103/105 Southampton Row, London WC1B 4HH, carries all these titles.

The IT Education Office in Rugby also has an Education Resource List, with materials that are free, apart from a contribution for post and packing.

Appendix 2 Curriculum Materials

Materials with a technological perspective

Longman are publishing, in collaboration with Nuffield, a major Design and Technology Project 11–16. This will incorporate some of the perspectives outlined in *Make the Future Work*.

To be published in 1993, containing student and teacher materials.

From the Centre for Alternative Technology

Project Resource Booklets on solar, wind and water power.

Teachers' Guides to Renewable Energy Projects (wind and water).

Do-it-Yourself plans on eight different alternative/appropriate technologies.

Information sheets – comprehensive introductions to various aspects of alternative technology.

Details from CAT (see Appendix 3 for address).

Other materials

The Greenhouse Effect

This highly visual pack, written by David Wright of the University of East Anglia, provides the teacher and pupils with a range of challenging yet manageable activities to help students understand the importance and nature of the greenhouse effect on our environment. The pack is aimed at the 11–14 age group, but can be used as an important part of GCSE classes too. It consists of a photocopiable resource book full of illustrations, comprehensive yet concise teachers' notes, a 30-minute VHS video and a large, full-colour poster.
WWF.

The Energy Project

This pack presents a novel approach to the theme of energy, encouraging pupils to challenge assumptions about energy use rather than just learning the technicalities of production and supply. Age group: middle/upper secondary. Copyright free, photocopiable student resource sheets include photos, cartoons, adverts, mapping, data analysis, carrying out science practicals, role-playing, etc. Concise Teachers' Notes give guidance for use of resources and checklists for expected outcomes. Suitable for the 11–16 age range,

materials have been designed to relate to Attainment Targets for Maths, Science and English, with two cross-curricular themes: Economic Awareness and Industrial Understanding and Environmental Education.
WWF.

Solving Problems: Design and Technology, Science and the Environment

Will be published in September 1992 by WWF (UK), in association with the DES.

Lots of ideas for the classroom and for in-service, drawing on case-study material collated through CREST and the WWF Environmental Enterprise Award Scheme.

Clean Water: A Right for All

Cross-curricular project book for 8–13 year-olds designed to help children discover the properties of water and its crucial role in their lives. Includes units on health, water supply and sanitation, pollution, disasters and the symbolic use of water in religion. Case studies from the UK and overseas.
1989, Unicef-UK.

SATIS (Science and Technology in Society), published by ASE, have several appropriate units.

Contact the Association for Science Education, College Lane, Hatfield, Herts AL10 9AA.

Global and development education

Earthrights: Education as if the Planet Really Mattered

Global Impact Project, York University. Ideas and activities for primary and secondary classrooms to help young people understand complex environmental and development issues, 1987.

World Studies 8–13: A Teacher's Handbook, by Simon Fisher and David Hicks

A practical resource for classroom teachers and for use on in-service courses. It contains instructions of over eighty classroom activities including many pages of copyright-free pupils' material to stimulate and develop children's interest in the wider world and to encourage them to view themselves and others in new ways. *Part 1*: Curriculum Planning; *Part 2*: Classroom Activities; *Part 3*: In-Service Ideas.
Oliver & Boyd.

Making Global Connections, edited by David Hicks and Miriam Steiner

A new resource book for teachers from the World Studies 8–13 Project providing practical teaching ideas and pupil activities on five issues of current global concern. Also looks at the theory and practice of world studies and links individual classroom practice with whole-school development.
1989, Oliver & Boyd.

Global Teacher, Global Learner, by Graham Pike and David Selby

Handbook for teachers exploring and developing theory and practice of global education as well as offering an extensive range of practical activities for the primary and secondary classroom. Important ideas and issues illustrated with cartoons, photographs and diagrams.
1988, Global Education Project, Hodder & Stoughton.

Teaching Development Issues. Manchester Development Education Project, 1986

Series of seven books which aim to support either teachers developing a unit as a topic on the 'Third World' or teachers introducing North/South development issues in a broader context. Structured to introduce a number of 'key issues', which allows teachers to extract material to suit their own work schedule. Copyright-free student stimulus material available. Ideal for GCSE.
Section 1: Perceptions
Section 2: Colonialism
Section 3: Food
Section 4: Health
Section 5: Population Changes
Section 6: Work
Section 7: Aid and Development

Rich World, Poor World, by Olivia Bennet

Raises some of the important issues in the Development debate: 'Is poverty caused by overpopulation?', 'Does food aid help the Third World?', and many more questions on disasters, trade employment aid, rural development, energy, education and appropriate technology. No easy answers given, difficult and often conflicting arguments presented. Ideal for GCSE.

What We Consume – Teachers' Handbook, and 10 pupil books

This pack provides a curriculum framework and classroom activities for teachers wishing to explore issues of environment and development with their pupils. One hundred original activities link pupils as consumers to economies and societies around the world. Age range: middle/upper secondary. Subject: cross-curricular. Each unit consists of a book of ten activities, including teachers' notes, pupils'

worksheets and a photoset of ten black and white pictures.
1990, WWF.

Population and Food, by Damian Randle

16-page, extensively illustrated booklet, dealing with hunger; food trade; food aid; self reliance; birth control; agribusiness and technology; involvement. Suggested discussion topics. GCSE.
1986, Arnold.

Choices in Development, Council for World Development Education

New series of six-sided A4 fold-out photo-illustrated sheets.
Sheet 1: Water, Water, Everywhere
Sheet 2: Aid for Development
Sheet 3: Trade and Development.
1989, CWDE.

The World Tomorrow – Hampshire Development Education Project 8–13

Twelve packs for Development Education topic work. Each pack contains two A5 booklets outlining practical approaches to curriculum planning to promote awareness of family, community and global interdependence. Key questions, starter activities, skills developed by activity and resources list are included in each pack.

Family Life	*Conservation*
Food for Thought	*Energy*
Celebrations	*Global Family*
Living Together	*On the Move*
Village Community	*A Developing Community*
A School Exchange	*A Teacher's Guide*

Town World Links

A teachers' pack produced by Manchester Development Education Project to help 13–16 year-olds undertake project work in their local area on themes demonstrating interdependence and community links with the wider world. Contains four A2 posters, Teachers' Book and five booklets of copyright-free student stimulus material. GCSE.

Doing Your Project	*The Media*
Communities	*Industry and Work*
Aid	

Choosing the Future

This pack aims, through an active student-centred learning approach, to introduce issues related to the use of natural resources in which the importance of balancing concepts of conservation and exploitation are evident. The topics include motorways, acid rain and tropical rainforests. Suitable for the middle/upper secondary range. Subjects covered include Economics, Geography and Science.

Reference books, video packs, software and simulations

Atlas of World Issues, by Nick Middleton

Lavishly illustrated with full-colour pictures, maps and diagrams. Information on current issues of global concern such as changing climate, overpopulation and urbanisation. 13–16 age range and GCSE. 1989, Hardback, OUP.

The Gaia Atlas of Planet Management

This book is a guide to a planet in critical transition. How humankind uses Earth's vast resources today will determine the health and ultimately the survival of our complex ecosphere for the decades and centuries to come. With a wealth of data, vivid graphics and authoritative text, the Atlas is an important resource for teachers and tomorrow's adults. 1985.

The Gaia Atlas of Future Worlds

This far-sighted and challenging book provides a toolkit for future choice, and offers hope – but hope dependent on a radical shift in human life style, perception and spirit.
1990, Robertson McCarta.

Atlas of the Environment

Over 200 full-colour maps and diagrams providing at-a-glance information on what is happening to our planet.
1991, WWF.

International Broadcasting Trust (IBT)

Developing Images/Images of the Developing World

28-minute VHS video and accompanying worksheets. Aims to help young people understand how television current affairs and documentary programmes are made and why particular images of the developing world predominate. Cross-curricular use: World Studies, Media Studies, Social Studies, Integrated Humanities, and Technology, where a global perspective is being introduced.
1988, IBT.

Choices for the Planet

Another IBT resource. Using excerpts from a range of environmental programmes, the pack contains a video, and teachers' notes and pupil activity sheets.
1990, IBT.

The Global Environment

BBC Education and IBT booklet, to accompany the ten 20-minute programmes broadcast each autumn. Contains students' activities and teachers' notes.
1990, IBT.

Links, by John Widdowson and Balbinder Rayt

Teachers' notes written to accompany 'Links', the series of five 23-minute school programmes which trace the economic relationships between rich and poor countries. Focus on UK, India and Tanzania. Aimed at 14–16 age range, GCSE. Well-illustrated with maps, diagrams and photographs. Programmes: Food Links; Colonial Links; Trade Links; Aid Links; Technology Links.
1988, IBT/BBC.

Computer software – Information Technology

SATCOM – WWF

This Information Technology computer project will encourage students to investigate global environmental issues. The satellite control computer-based (SATCOM) learning package gives in-depth treatment to nine environmental issues, including ozone depletion, deforestation, acid rain and urbanisation. These issues are set in a range of nine named geographical areas.

It has a wide range of different data elements, and is supported by a manual of resources and pupil activities.
1991, WWF.

The Water Game – Centre for World Development Education

A program based on the daily use and supply of water. It is a simulation in which the user (with friends) goes to a waterless country cottage, and has to estimate the water needs of the humans, animals and crops. The pack consists of one disc, a 32-page booklet and a wide range of useful resource material.
1990, CWDE.

Sand Harvest – CWDE

A role-play simulation of desertification in the Sahel, based on life in Mali. The users adopt different roles (nomad, government officer and villager) and make decisions which affect their survival. The pack consists of four booklets, role cards, two wall charts and one disc.
1990, CWDE.

Role-play/games

The Trading Game

By taking part in this simulation activity pupils experience how trading relations can affect a country's prosperity. Working as groups, the pupils represent rich and poor nations. During the simulation pupils must trade technology and raw materials in order to manufacture different products. Originally written for older pupils, this activity can be adapted for use at Key Stage Three.
1990, Christian Aid, P. O. Box 100, London SE1 7RT

Development without Destruction

A board game published by Action Aid – based on Snakes and Ladders, but substituting trees and chimneys. Aims to raise awareness and explore the themes of culture, environment, cooperation and development. For 4–5 players at a time. Available in four languages.
1991.

Full catalogues are available from CWDE, WWF, Christian Aid, OXFAM, and Action Aid – addresses in Appendix 3.

Background reading for teachers

Poverty and the Planet: A Question of Survival, by Ben Jackson
To the Majority World,environmentalism is a diversion from the real problems of economic injustice and exploitation. This book rejects the short-term development projects that destroy the resources of the poor, and calls for a new ecological development.
210 pp., 1990, WDM/Penguin.

Renewing the Earth: Development for a Sustainable Future, by Seamus Cleary

This book explores the connection between conventional development theory and what has happened in practice, reflecting on the missed opportunities of forty years of development assistance to the Third World and its effects on the environment.
148 pp., 1989, CAFOD.

Small is Beautiful: A Study of Economics as if People Mattered, by E.F. Schumacher

Small is Beautiful is a revolutionary book. It proposes a system, using intermediate, or appropriate, technologies, with local labour resources, with the emphasis on the person, not the product. The author stresses the need to return to wisdom in planning the future.
226 pp., 1974, Abacus.

Greenprints for Changing Schools, by Sue Greig, Graham Pike and David Selby

Handbook for decision makers in the education system. Links theory and practice of educational change and looks at strategies for integrating environment and development concerns into everyday classroom teaching.
1989, WWF/Kogan Page.

Teaching Green, by Damian Randle

Outlines the agenda for the greening of the curriculum, and provides lots of examples of green education in action.
1989.

Inside the Third World, by Paul Harrison

First-hand paperback account of realities of life for the poor in Asia, Africa and Latin America.
Pelican, 2nd ed, 1981, reprinted, revised overview, 1987.

The Greening of Africa, by Paul Harrison

International Institute for Environment and Development/Earthscan study tracing the roots of Africa's environment and development problems. Surveys ventures in sustainable development schemes that have mobilised peasants to boost food production and conserve the environment.
1987, Paladin.

The Third World Tomorrow, by Paul Harrison

Very readable paperback with first-hand reports of how 'self-help', small-scale appropriate technologies, and other concepts are being put into practice in Asia, Africa and Latin America.
1980, Pelican, 2nd ed, 1983.

How the Other Half Dies, by Susan George

Provocative and readable study of international food problems, how the rich aren't helping.
1976, Penguin, reprinted with foreword, 1986.

Towards Sustainable Development

Explores the concept of sustainability through 14 case studies of grass-roots development prepared by independent African and Asian journalists.
1987, Panos.

Defending the Future: a Guide to Sustainable Development, by Johan Holmberg, Stephen Bass and Lloyd Timberlake

This 40-page A4 booklet examines the issues of liberty, equality, justice...and sustainable development, by looking at the current state of the world: money, power and institutions, and concludes by setting out the common North/South agenda.
1991, Earthscan Publications/International Institute for Environment and Development.

Global Consumer, published jointly by *New Consumer* and *Gollancz*, is a consumer guide to the products that represent best value with the Majority World in mind. The book is designed to be the first popular introduction to the major trade and development issues that affect the Majority World. It is highly readable, and covers a wide range of consumer goods – from DIY to fruit juice and holidays – and would be a useful resource for teachers tackling Economic and Industrial Understanding and Business Studies.
1991.

Africa in Crisis: The Causes, the Cures of Environmental Bankruptcy

An Earthscan paperback which analyses the causes of Africa's famines, showing how misplaced priorities have turned droughts into major disasters. Ecological, political and economic factors are all carefully investigated in this informative book.

Developed to Death: Rethinking Third World Development, by Ted Trainer

A devastating yet constructive critique of what today's economists and politicians say about development. Trainer shows that development must be within the demand of global sustainability, and that the rich are getting richer because the poor are getting poorer. He argues that there are alternatives which would yield a just, peaceful and sustainable world.
1989, Green Print.

All the books in this list of resources are available from the IT Bookshop, 103–105 Southampton Row, WC1B 4HH

Appendix 3 Useful Addresses

Action Aid
Hamlyn House
MacDonald Road
London N19 5PG
071 281 4101

Birmingham DEC
Gillett Centre
Selly Oak Colleges
Birmingham B29 6LE
021 472 3255

Catholic Fund for Overseas
Development (CAFOD)
2 Romero Close
Stockwell Road
London SW9 9TY
071 733 7900

Centre for Alternative Technology
Machynlleth
Powys
Wales SY20 9AZ
0654 703743

Centre for Global Education
University of York
Heslington
York YO1 5DD
0904 433 4444

Centre for World Development
Education
1 Catton Street
London WC1R 4AB
071 831 3844

Christian Aid
PO Box 100
London SE1 7RT
071 620 4444

Commonwealth Institute
Kensington High Street
London W8 6NQ
071 603 4535

Council for Education in World
Citizenship (CEWC)
Seymour Mews House
Seymour Mews
London W1H 9PE
071 935 1752

Council for Environmental
Education
Faculty of Education
University of Reading
London Road
Reading RG1 5AQ
0734 756061

Global Futures Project
10 Woburn Square
London WC1H ONS
071 636 1500

Intermediate Technology (IT)
Myson House
Railway Terrace
Rugby CV21 3HT
0788 560631

Intermediate Technology
Publications/Bookshop
103–105 Southampton Row
London WC1B 4HH
071 436 9761

International Broadcasting
Trust (IBT)
2 Ferdinand Place
London NW1 8EE
071 482 2847

Manchester Development
Education Project
Manchester Polytechnic
801 Wilmslow Road
Manchester M20 8RG
061 445 2495

National Association of
Development Centres (NADEC)
6 Endsleigh Street
London WC1H 0DX
071 388 2670

New Economics Foundation
88–94 Wentworth Street
London E1 7SE
071 377 5696

Oxfam
274 Banbury Road
Oxford OX2 7DZ
0865 311311

Passe Partout
72 St John's Street
London EC1M 4DT
071 251 0074

Save the Children Fund
Mary Datchelor House
17 Grove Lane
London SE5 8RD
071 703 5400

Scottish Catholic International
Aid Fund (SCIAF)
5 Oswald Street
Glasgow G1 4QR
041 221 4447

Tourism Concern
Froebel College
Roehampton Lane
London SW15 5PU
081 876 2242

Traidcraft
Team Valley Trading Estate
Gateshead NE11 ONE
091 491 0591

UNICEF
55–56 Lincoln's Inn Fields
London WC2A 3NB
071 405 5592

Voluntary Service Overseas (VSO)
317–325 Putney Bridge Road
London SW15 2PN
081 780 2266

WaterAid
1 Queen Anne's Gate
London SW1H 9BT
071 222 8111

World Development Movement
(WDM)
25 Beehive Place
London SW9 7QR
071 737 6215

World Wide Fund for Nature (WWF)
Panda House
Weyside Park
Godalming GU7 1XR
0483 426444

Appendix 4   Intermediate Technology

Intermediate Technology is an international development agency, founded in 1965 by Dr E.F. Schumacher.

IT's main task is to work through local organisations in certain countries overseas to develop appropriate technologies and techniques to enable poor producers to become more productive. This enables people to earn a better living for themselves and their communities and is part of the process of sustainable development, as it is IT's aim to decrease dependency and increase self-sufficiency.

The guiding principle is to find the way to make the best use of local ingenuity, skills and resources, listening to what the local people say, and responding to the local context. Much of the work is carried out with women, on whom much of the burden of feeding, clothing and caring for the family falls.

IT concentrates its work in: Bangladesh, Kenya, Sri Lanka, Sudan, Peru and Zimbabwe. There is also programme work in South India, Nepal and Malawi, and IT has a wholly owned subsidiary company which carries out consultancies in countries other than those listed above.

The charity has particular areas of expertise:

- *Agriculture and Fisheries* – fishing, animal health care and agriculture;
- *Rural manufacturing* – carpentry and blacksmith tools, rural transport, 'micro' hydro-electricity generation, and textiles;
- *Agro processing* – small-scale food processing and fuel-efficient stove manufacture;
- *Minerals and shelter* – small-scale mining and housing material manufacture.

IT also has a publications company, IT Publications, which publishes about thirty books a year on development, technical and related issues, both for specialists and general interest. These books are marketed all over the world, and form an important part of the communication process.

IT employs about 200 people world-wide, with more than half of its staff in and from the countries within which IT concentrates its work. There are technical specialists, social scientists, experts on training, communications and business: IT can be said to work in a truly multi-disciplinary way.

In the United Kingdom lies the responsibility for communicating the issues and the realities to decision makers and the UK public, which includes teachers and their pupils. That is the task of the IT Education Office, which uses the knowledge and experience from IT's work, and converts it into a form appropriate for the Curriculum.

Acknowledgements

We are grateful to the following for permission to reproduce copyright material:

Catholic Fund for Overseas Development for an abridged extract by Bernard Guri from *Link* magazine, Autumn 1989; Earthscan Publications Ltd for an extract & an abridged extract from *Blueprint for a Green Economy* by David Pearce (1989), © Crown Copyright; Faber & Faber Ltd for the poem 'Going, Going' from *High Windows* by Philip Larkin; National Curriculum Council for an abridged extract from *Curriculum Guidance 7: Environmental Education*; Zed Books Ltd, London, for an abridged extract from *Staying Alive: Women, Ecology and Development* by Vandana Shive (1988) & an abridged extract from *Abandon Affluence* by Ted Trainer (1985).

We have been unable to trace the copyright holder in 'A Parable' by Jon Rye Kinghorn from *Global Teacher, Global Learner* by Graham Pike & David Selby (Hodder & Stoughton, 1988) & would appreciate any information that would enable us to do so.

We are grateful to the following for permission to reproduce photographs and other illustrations:

Sophie Baker, page 16; Basement Art, Shadwell Centre, pages 131, 135 (left and right); John Birdsall Photography, page 107; Karen Coulthard, Headteacher, Berger Junior School, page 97 (Fig. 7.3); Martin Downie, pages 125, 128; Mark Edwards/Still Pictures, pages 168, 170; Intermediate Technology, pages 34, 39, 58, 59, 95, 109 (top) (photo Jeremy Hartley), 112 (top and bottom), 121, 136, 149, 152 (top and bottom), 155, 177; Intermediate Technology Publications, *Stove Maker, Stove User*, Teacher's Notes, page 124 (Fig. 9.2); *Small World*, page 110 (Figs. 8.4 and 8.5); L. Taylor and P. Jenkins, *Time to Listen*, page 100 (Fig. 7.7); Leeds Postcards (photo Pete Betts, 1990), page 56; Manchester Evening News, page 109 (bottom); Mehfil-E-Tar Bedford Asian Women's Textile Group, page 141; Colin Mulberg, pages 30, 31; Open University Press, John Elliott, *Action Research for Educational Change*, page 73 (Fig. 5.4); Roger Standen, pages 81, 82, 85 (left and right), 86; Unicef, pages 43 (Fig. 3.3) (photo Vivianne Holbrooke), 50 (Fig. 3.10), 138, 139 (5 photos), 140; Volvo Concessionaires, page 40.

We are unable to trace the copyright holders of the following: page 37 and page 98 and would appreciate any information enabling us to do so.

Illustrations by Michael Salter.